T0340067

The Hydrophiloidea
(Coleoptera)
of Fennoscandia and Denmark

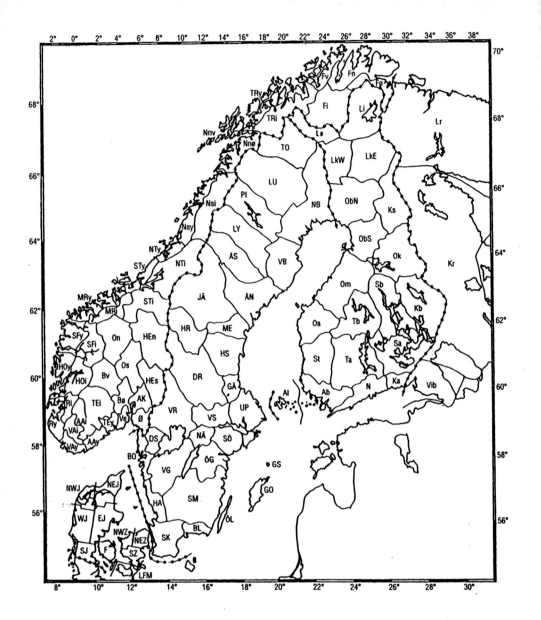

List of abbreviations for the provinces used throughout the text, on the map and in the following tables.

DENMARK

SJ	South Jutland	LFM	Lolland, Falster, Møn
EJ	East Jutland	SZ	South Zealand
WJ	West Jutland	NWZ	North West Zealand
NWJ	North West Jutland	NEZ	North East Zealand
NEJ	North East Jutland	B	Bornholm
F	Funen		

SWEDEN

Sk.	Skåne	Vrm.	Värmland
Bl.	Blekinge	Dlr.	Dalarna
Hall.	Halland	Gstr.	Gästrikland
Sm.	Småland	Hls.	Hälsingland
Öl.	Öland	Med.	Medelpad
Gtl.	Gotland	Hrj.	Härjedalen
G. Sand.	Gotska Sandön	Jmt.	Jämtland
Ög.	Östergötland	Ång.	Ångermanland
Vg.	Västergötland	Vb.	Västerbotten
Boh.	Bohuslän	Nb.	Norrbotten
Dlsl.	Dalsland	Ås. Lpm.	Åsele Lappmark
Nrk.	Närke	Ly. Lpm.	Lycksele Lappmark
Sdm.	Södermanland	P. Lpm.	Pite Lappmark
Upl.	Uppland	Lu. Lpm.	Lule Lappmark
Vstm.	Västmanland	T. Lpm.	Torne Lappmark

NORWAY

Ø	Østfold	HO	Hordaland
AK	Akershus	SF	Sogn og Fjordane
HE	Hedmark	MR	Møre og Romsdal
O	Opland	ST	Sør-Trøndelag
B	Buskerud	NT	Nord-Trøndelag
VE	Vestfold	Ns	southern Nordland
TE	Telemark	Nn	northern Nordland
AA	Aust-Agder	TR	Troms
VA	Vest-Agder	F	Finnmark
R	Rogaland		

n northern   s southern   ø eastern   v western   y outer   i inner

FINLAND

Al	Alandia	Kb	Karelia borealis
Ab	Regio aboensis	Om	Ostrobottnia media
N	Nylandia	Ok	Ostrobottnia kajanensis
Ka	Karelia australis	ObS	Ostrobottnia borealis, S part
St	Satakunta	ObN	Ostrobottnia borealis, N part
Ta	Tavastia australis	Ks	Kuusamo
Sa	Savonia australis	LkW	Lapponia kemensis, W part
Oa	Ostrobottnia australis	LkE	Lapponia kemensis, E part
Tb	Tavastia borealis	Li	Lapponia inarensis
Sb	Savonia borealis	Le	Lapponia enontekiensis

USSR

Vib   Regio Viburgensis   Kr   Karelia rossica   Lr   Lapponia rossica

FAUNA ENTOMOLOGICA SCANDINAVICA
Volume 18   1987

The Hydrophiloidea
(Coleoptera)
of Fennoscandia and Denmark

by

Michael Hansen

E. J. Brill/Scandinavian Science Press Ltd.

Leiden · Copenhagen

Fauna entomologica scandinavica
is edited by "Societas entomologica scandinavica"

Editorial board
Nils M. Andersen, Karl-Johan Hedqvist, Hans Kauri,
N. P. Kristensen, Harry Krogerus, Leif Lyneborg,
Hans Silfverberg

Managing editor
Leif Lyneborg

World list abbreviation
Fauna ent. scand.

Colour reproduction
Cromoscan, Rødovre, Denmark

Composition and printing
Vinderup Bogtrykkeri A/S
7830 Vinderup, Denmark

ISBN 90 04 081836
ISBN 87-87491-35-4
ISSN 0106-8377

Author's address:
Zoological Museum
Universitetsparken 15
DK-2100 Copenhagen
Denmark

Contents

Plates 1-4 are arranged after p. 224.

Abstract

The Hydrophiloidea, comprising the families Hydraenidae, Spercheidae, Hydrochidae, Georissidae, and Hydrophilidae, of Denmark and Fennoscandia are monographed. The 118 species hitherto known from the region, and 18 additional species from the adjacent areas, are included. All taxa are keyed, described, and illustrated. Notes on the biology and distribution are given. *Ochthebius kaninensis* Poppius, 1909 is considered a valid species, distinct from *O. bicolon* Germar, 1824. The name *Pseudenochrus* Lomnicki, 1911 is considered a synonym of *Lumetus* Zaitzev, 1908. Type-species are designated for the following names: *Hymenodes* Mulsant, 1844: Type-species: *Ochthebius pellucidus* Mulsant, 1844 (= *O. nanus* Stephens, 1829). *Bilimneus* Rey, 1883: Type-species: *Hydrophilus atomus* Duftschmid, 1805. *Enoplurus* Hope, 1838: Type-species: *Hydrophilus spinosus* Steven, 1808.

Material studied and state of knowledge

Though the Coleoptera are regarded one of the more well studied insect orders, the hydrophiloids are a neglected group in many respects. They have never been paid much attention by collectors, and few specialists have dealt with them. This has resulted in a much less satisfactory knowledge of their systematics and faunistics than is the case with many other families.

So, though the material deposited in the Scandinavian museums and in the private collections may seem considerable, it should not be considered satisfactory (except perhaps for Denmark and southern Sweden). This is illustrated by the fact that many species are indeed very common and widespread, but are unrecorded from provinces where they no doubt occur. Thus many of the provincial records are based on but a single or a few specimens, and if detailed maps are made, many species show only very sparse records. This becomes less obvious when larger districts are considered (as in the present work) and this scale of recording probably makes a comparison between the different species more reliable, as it compensates to some extent for the low collecting intensity.

The numbers of species recorded from each of the provinces in the treated area is mapped in Fig. 1. Due to the fact that both the size of the provinces and the collecting intensity vary considerably, the map should be interpreted with some reservation. There are many procinces – the northern ones in particular, and the Norwegian ones in general – where collecting intensity has been remarkably low; this means that the numbers are not directly comparable. Taking this into account, the map in fact only reflects the general situation: a decrease in the number of species towards the Northwest. Other conclusions ought not to be based on the material available at present.

Most of the records given in this work were mentioned by Lindroth (1960), but about 15 species of hydrophiloids are recorded as new to Fennoscandia and Denmark since then, and many additional provincial records are given for the "old" species. Further, some changes concerning the division into provinces have taken place. Primarily, a further division of Denmark into 11 provinces (instead of 3) made it necessary to revise

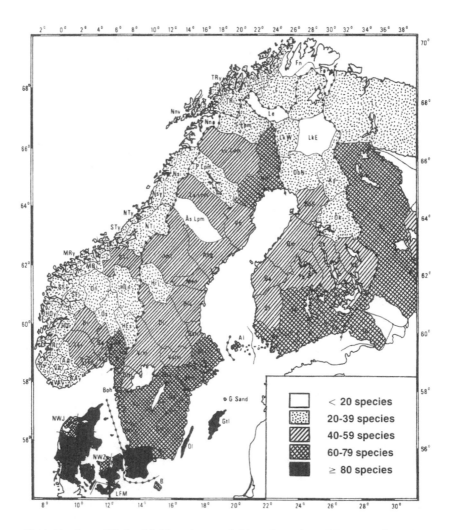

Fig. 1. Numbers of Hydrophiloid species recorded in each province of Fennoscandia and Denmark.

all available Danish material, and this resulted in the discovery of some species not previously known from Denmark (or even Scandinavia). Also, there have been some changes concerning the East Fennoscandian provinces, being mainly a gathering of the Russian provinces into fewer, large provinces.

From Sweden, Finland and (to a lesser extent) Norway a large number of specimens have been examined. Especially in the more difficult groups (e.g. *Helophorus*) I have made efforts to check as many records as possible. For the more characteristic species, the records are not so consistently checked.

These examinations have resulted in many corrections and additions to Lindroth's catalogue. Where these changes may be of interest (i.e. deletion of records, or taxonomic re-interpretations), notes are given under the respective species.

Morphology of the adult

Head

Labrum well developed, normally visible in dorsal view. Clypeus large, its lateral margins concealing the antennal base to some extent (dorsal view). Clypeus separated from frons by a transverse, normally V-shaped suture or furrow, that often continues posteriorly in a longitudinal median frontal suture or furrow; these furrows (or sutures) are almost always distinct; only in a few clearly derived forms (eg. *Chaetarthria*) have they disappeared completely. The interpretation of the cephalic components has been the subject of some disagreements between authors (the portion that is here termed clypeus, has elsewhere been referred to as frons, or frons + clypeus ("frontoclypeus")). The terms used here are merely to be regarded as trivial names; they are in accordance with e.g. Hansen (1931), Lohse (1971), and Smetana (1978).

The compound eyes are well developed, often rather convex, posteriorly reaching the anterior margin of the pronotum (or almost so). The head is narrowed just behind the eyes, often rather abruptly so.

Mentum large, rectangular or trapezoid (narrowed anteriorly); gula normally distinct, delimited by two well separated gular sutures. The ventral face of the head – except mentum and the maxillae – with fine rugulose, reticulate or punctate microsculpture, and fine and dense hydrofuge pubescence, which extends laterally covering also the temporae.

Labium small and weakly sclerotized, bearing a pair of small 3-segmented labial palpi.

Maxillae well developed, with strongly prolonged (only seldom shorter) 4-segmented maxillary palpi; the palpi normally distinctly longer than the antennae, their basal segment minute.

The antennae (e.g. Figs 96, 304) short, inserted just anterior to the compound eyes, below the expanded clypeal margin, so the antennal base is concealed in dorsal view (except in *Cercyon* and related genera, where the clypeal margin is excised at the antennal base). The number of antennal segments varies from 7 to 11 (in European genera

11

with no more than 9 segments). 1st segment (the scape) markedly prolonged, the 2nd (the pedicel) also enlarged, but normally shorter than 1st, distal to the pedicel with 1 to 3 small and uniform segments, followed by a larger, more or less cup-shaped segment (the cupule), bearing a 3- or 5-segmented club. The club is provided with fine and dense hydrofuge pubescence, the segments proximal to it are glabrous (except in *Spercheus*, Fig. 117, in which also the pedicel and the cupule are densely pubescent). When at rest, the antennae are directed posteriorly, with the basal segments lying closely up against the ventral margin of the compound eyes, and the club being received in a variably marked antennal excavation at the antero-lateral portions of prosternum.

Thorax

Shape of prothorax variable, often much wider than long, normally narrowed anteriorly, in many species also narrowed posteriorly. Surface of the dorsal portion, pronotum, rather uneven in some genera, with distinct furrows or depressions (in genera

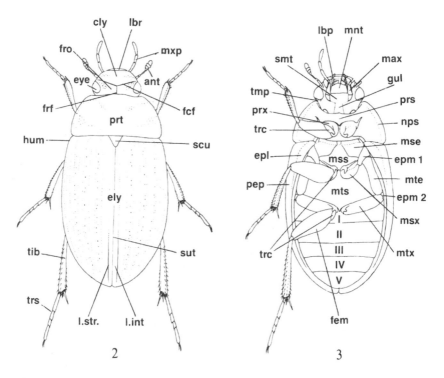

Figs 2-3. General structure of a hydrophiloid. – 2: dorsal surface; 3: ventral surface. For abbreviations: see p. 23.

12

with the pronotum narrowed posteriorly); quite even in other genera. Surface of the ventral portion, prosternum, normally with fine granulose, reticulate or punctate microsculpture and dense hydrofuge pubescence, except near the lateral margins, which are more or less narrowly shining and glabrous, usually well demarcated from the pubescent portion by a more or less distinct notopleural suture. Lateral portions of prosternum (inside the notopleural suture) often with distinct antennal excavations (see e.g. Fig. 219). Prosternum (e.g. Figs 43, 68) with well developed coxal cavities, that are normally separated by a raised prosternal process; prosternum anterior to the coxal cavities short and wide, often raised medially; sometimes (*Georissus*, Fig. 134) the prosternum is reduced to a narrow chitinous bridge lacking any trace of the median prosternal process, completely covered by the procoxae (and -trochanters), and having no procoxal cavities (except laterally).

Scutellum generally small, but normally visible.

Mesosternum normally trapezoid, strongly narrowed anteriorly, often raised in middle, forming a longitudinal ridge, a posteromedian process, a well marked plate, etc.

Metasternum large, rectangular, or slightly trapezoid (narrowed anteriorly) with middle portion normally distinctly raised (but much more bluntly than mesosternum); the middle metasternal portion sometimes demarcated from the lateral portions by fine oblique lines (femoral lines) (see e.g. Fig. 261).

The surface of meso- and metasternum with episternae and epimeres (as the prosternum) normally with fine microsculpture and fine and dense hydrofuge pubescence; often the raised middle portions are glabrous or only sparsely pubescent; but only seldom (eg. *Georissus, Megasternum*) completely without pubescence.

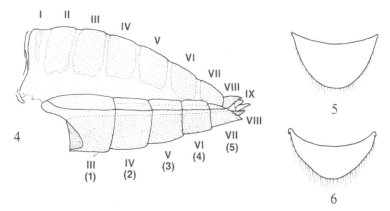

Figs 4-6. Abdomen and abdominal segments in hydrophilids *(Helophorus)*. – 4: abdomen in lateral view; 5: 8th tergite; 6: 8th sternite.

Abdomen

First two abdominal tergites without corresponding sternites (Fig. 4); the first tergite incomplete, represented only by two separate sclerotized plates. In the Hydraenids there are 10 distinct tergites, while in the other families only 9 tergites can be recognized. However the 9th tergite in males of Hydrophilidae shows some division (Fig. 9) and may correspond to 9th + 10th tergite in the Hydraenids. It is possible that a 10th tergite in females of Hydrophilidae is part of the modified sclerites, that make up their copulatory organ.

The number of visible abdominal sternites varies from 5 to 7 (1st visible sternite representing abdominal segment 3). In the Hydraenids where the number of visible sternites is 7 or (in some males) 6 the sternite and tergite of segment 8 do not enclose the sclerites of the following segments, whereas this is the case in the other families where the number of visible sternites is usually 5. Apparently no 10th sternite is present (except that in females of the Hydrophilidae it may be part of the modified sclerites making up the copulatory organ). The genital segment (urite 9) is further described in combination with the copulatory organ.

The visible abdominal sternites are normally (as are the thoracic sternites) provided with fine rugulose, reticulate or punctate microsculpture, and fine and dense hydrofuge pubescence, which may however be partially absent (e.g. Hydraenidae, *Hydrophilus, Laccobius*) or even lacking (e.g. *Georissus, Megasternum*).

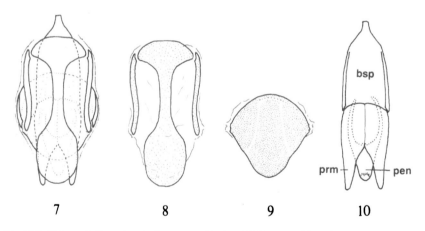

Figs 7-10. Male genitalia and genital segment of hydrophilids. – 7: position of aedeagus and genital segment (ventral view); 8: 9th sternite; 9: 9th tergite; 10: aedeagus (dorsal view).

14

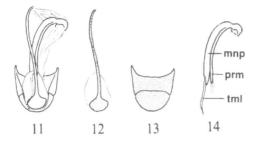

Figs 11-14. Male genitalia and genital segment of hydraenids. – 11: position of aedeagus and genital segment (ventral view); 12: 9th sternite; 13: 9th + 10th tergite; 14: aedeagus.

Male genitalia

The male copulatory organ, the aedeagus, is of great taxonomic importance, and may be very helpful in distinguishing closely related species. There appear to be diverging opinions as to the nomenclature of the components of the genitalia; the terms used here follow Lindroth (1957); it is also termed the aedoeagus. English authors refer to it as the aedeagophore (and term the penis (sensu Lindroth, 1957) the aedeagus). Two main types of male genitalia can be recognized within the Hydrophiloidea:

In Spercheidae, Hydrochidae, Georissidae and Hydrophilidae the aedeagus (Fig. 10) is of the trilobed type, consisting of a basal piece, two lateral lobes (parameres) and a median lobe (penis). It normally rests in the abdomen in a primitive position (Fig. 7), but may be rotated along its axis (*Berosus*) so it lies on its side in an oblique position in the abdomen. The basal piece is usually weakly sclerotized or membranous dorsally, and forms together with the parameres a tube in which the penis slides. The penis is prolonged basally into two lateral, often rod-like struts. The whole aedeagus is normally flattened dorso-ventrally.

The other type of aedeagus is found in the Hydraenidae. It consists basically of a well sclerotized tubular main piece, which is strongly dorso-ventrally curved at base, rotated along its axis, and lying on its side in an oblique position in the abdomen (Figs 11, 14). It usually possess two well developed parameres inserted (laterally or ventrally) near its base, and terminates in a movable (usually asymmetrical) lobe. In some Hydraenids (e.g. some *Limnebius*) the components of the aedeagus are strongly asymmetrical, forming a very complex structure in which the components can be difficult to recognize. In other Hydraenids there tends to be a reduction of the aedeagal components, and in some forms the parameres are completely missing. It is possible that the tubular main piece is homologous to the basal piece of the trilobed type, and the terminal lobe to the penis (Perkins, 1980).

The sternite of 9th abdominal segment (genital segment) has a median sclerite (Figs 8, 12), which is usually wide apically, then strongly narrowed and (except in Hydraenidae) widened again basally; sometimes (e.g. *Cercyon*) there is no distinct apical widening. Two narrow elongate lateral sclerites articulate latero-basally to the median sclerite (except in Hydraenidae).

Perkins (1980) suggests that the wide apical portion of the (median) sclerite in the Hydraenids represents a 10th sternite, due to the presence of 10 distinct tergites.

15

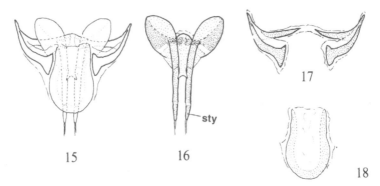

Figs 15-18. Sclerotized parts of female genitalia of hydrophilids. – 15: position of the modified sclerites and the genital segment; 16: same, but with sclerites of genital segment removed; 17: 9th sternite; 18: 9th tergite.

Female genitalia

Contrary to the male genitalia, the female genitalia are of only little value in determining the species. The spermatheca, which in many other groups of Coleoptera provides good distinguishing characters, is normally small and membranous in the Hydrophiloidea, though it may be more sclerotized in some forms (some Hydraenidae, some *Laccobius*). It is rather variable, but normally very alike in closely related species.

In the Hydraenidae the sternite and tergite of 9th abdominal segment are simple and similar to the preceding sclerites, and are not retracted into the abdomen.

In the other families they are enclosed between the 8th sternite and tergite, and are strongly modified (Figs 15-18). The 9th sternite is divided medially, and laterally bent to the dorsal face, where it articulates to the weakly sclerotized 9th tergite. Enclosed between the 9th sternite and tergite is a complex of strongly modified sclerites, that terminate in a pair of styli (Fig. 16), functioning in the formation of the egg cocoon (Maillard, 1968).

Figs 19-21. Last abdominal sclerites in female hydraenids. – 19: abdominal apex (in oblique postero-dorsal view); 20: 8th sternite; 21: 9th sternite.

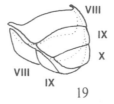

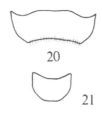

Elytra

The elytra are well developed, completely covering the abdomen (except sometimes for the very apex). They are of rather variable shape, mostly with distinct longitudinal striae or series of punctures; the number of striae (or series) is normally 10. Only in a few genera are there deviations from this number; thus in many *Hydraena*-spp. there may be 12-15 striae, and in most *Laccobius*-spp. the number is about 20; most likely the increase in number is a derived condition.

In some forms (eg. Hydrophilinae and many Hydrobiinae) such series are obsolete or completely absent, but they may normally be traced easily on the ventral face of the elytra.

In some species (*Hydrobius,* some *Ochthebius,* some *Helophorus,* etc.) a short intercalary stria or row of punctures is present at the elytral base between 1st and 2nd stria. (The striae are numbered from the elytral suture towards the lateral margin, so the 1st stria is the one nearest to the suture; it is also referred to as the sutural stria.).

The elytral interstices are numbered in the same way as the elytral striae, so the 1st interstice is that between the 1st stria and the suture. In a few genera (e.g. *Hydrophilus, Hydrochara, Hydrobius*) the alternate interstices (i.e. 3, 5, 7, 9 and 11) are provided with longitudinal rows of coarse setiferous punctures, that may sometimes be very conspicuous; such setiferous punctures may also be found in some genera without punctato-striate elytra (e.g. some *Enochrus*).

The epipleural portion of the elytra appears to be unique; it consists of an inner and an outer portion (e.g. Fig. 3). Only the inner portion represents the epipleuron; the outer portion is the ventral face of the (sharply ridged) outermost elytral interstice, and is here referred to as the pseudepipleuron. In some genera (e.g. *Ochthebius, Helophorus*) the pseudepipleuron is delimited from the true epipleuron by a fine sharp longitudinal ridge, the former being glabrous, the latter pubescent; in others (e.g. *Hydrobius*) the pseudepipleuron is rather vertical (and glabrous), the epipleuron rather horizontal (and finely pubescent); in *Cercyon* the pseudepipleuron and the epipleuron are quite uniform (glabrous), being only separated by a very fine suture.

The ventral surface of the elytra has a more or less distinct, slightly duller field of microscopical stridulatory spines, laterally on the level of the 1st visible abdominal sternite. Though it is rather variable in shape and appearance, such a stridulatory field seems to be generally present within the Hydrophiloidea.

Hind wings

The hind wings are normally well developed, but may in some forms (e.g. some *Hydraena, Georissus)* be rudimentary. They may very seldom even be completely missing (*Hydraena,* cf. Balfour-Browne, 1958). Wing dimorphism is very rare and apparently known only in *Helophorus granularis.*

The Hydrophiloidea show two different main types of wing venation: in Spercheidae, Hydrochidae, Georissidae and Hydrophilidae, the cubitus and media form a loop, from which a single vein continues towards the edge of the wing (cantharoid

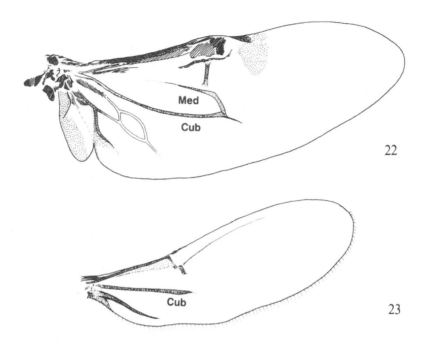

Figs 22-23. Right hind wing of hydrophiloids. – 22: *Laccobius;* 23: *Ochthebius.*

type, Fig. 22). In the Hydraenidae the venation is strongly reduced and there is no medial-cubital loop (staphylinoid type); of the above mentioned veins only the cubitus can be recognized (Fig. 23).

Legs

The legs are variable, sometimes rather long and slender, sometimes short and stout, in general developed for climbing rather than for swimming.

Procoxae more or less spherical, sometimes a little transverse, only narrowly separated, often almost touching each other. Mesocoxae usually transverse, and normally narrowly separated. Metacoxae strongly transverse, reaching almost to the lateral margins of the body. All coxae normally with fine and dense hydrofuge pubescence (except in species with glabrous thoracic and abdominal sternites). The trochanters are normally distinct.

Femora often with a well defined basal field of dense hydrofuge pubescence (the distal portion, or at least the apex glabrous).

Tibiae with longitudinal series of fine or very fine (seldom strong) spines, and occasionally (*Berosus*) with a fringe of long fine swimming hairs on the meso- and metatibiae.

18

Tarsi usually 5-segmented; the basal segment minute and sometimes not visible in dorsal view. In some genera this basal segment has disappeared, so the tarsi are 4-segmented, e.g. *Georissus* (all tarsi), *Cymbiodyta* (meso- and metatarsi) and *Berosus* (male protarsi). In other genera the 3 basal segments are small and closely articulated, so the tarsi appear 3-segmented (e.g. *Hydraena, Limnebius*). Only the Sphaeridiini have 5-segmented tarsi with long basal segment.

All tarsi with two claws.

Sexual dimorphism

In general, the hydrophiloids do not show much sexual dimorphism. In some genera there are no secondary sexual characters, and in genera where such are present, the differences are mostly rather inconspicuous.

The commonest type of dimorphism is shown by the protarsi. They are often slightly dilated in males, especially the basal protarsal segments (*Limnebius,* some *Ochthebius, Berosus, Laccobius*); sometimes they are more markedly dilated distally (*Sphaeridium, Hydrophilus*) and often with stronger or more curved protarsal claws.

The abdominal sternites sometimes provide good secondary sexual characters (e.g. in *Limnebius,* where males have an enlarged 6th sternite, or in some *Hydraena,* where the number of visible sternites may distinguish the sexes).

In some Sphaeridiinae (e.g. *Cercyon*) males are characterized by a sucking-disc shaped plate on the maxillae.

More types of sexual dimorphism are found, but they are on a much more specific level, and are described under the respective species.

Larval characteristics

The Hydrophiloid larvae show a great morphological diversity, varying considerably in body shape (Figs 31-36), and there seems to be no single character holding the forms together.

Head well sclerotized, generally projecting freely, somewhat inclined (Hydraenidae), almost horizontal (Spercheidae, Hydrochidae), or somewhat elevated (most Hydrophilidae). As usual in coleopterous larvae, it has a distinctive pattern of sutures, normally forming a V-or Y-shaped suture. In Hydraenidae (Figs 24, 25) labrum is movable, not fused with the clypeus. In other hydrophiloids (Figs 27,28) labrum and clypeus have fused, forming together a nasale, which is often crenate and may be asymmetrical; each side of the nasale has a projecting lobe (epistomal lobe); right and left lobes may be alike or differently shaped, often provided with stiff marginal setae.

Eyes usually composed of 5 or 6 ocelli, that may be well separated or (e.g. *Cercyon)* very close-set. Rarely with only 3 ocelli on each side of head (*Sphaeridium*).

Antennae normally inserted on dorsal surface of head, 3-segmented. 2nd segment (beside terminal segment) with a small joint-like appendage.

Mandibles variable, right and left mandible alike, or – often – differently shaped. In Hydraenidae (Fig. 26) they have a large basal molar area and a well developed movable appendage (prostheca) inserted on the inner face. In other hydrophiloids there is no prostheca and (except for Hydrochidae) no molar area.

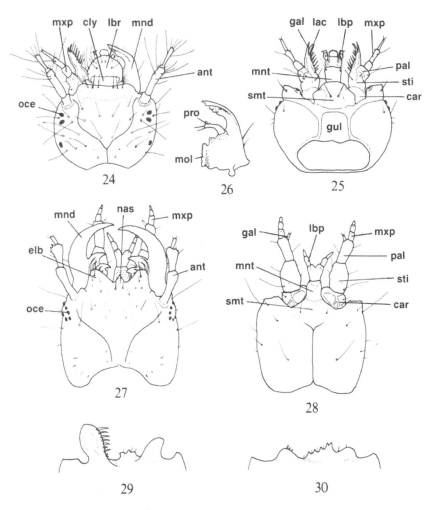

Figs 24-30. Head of hydrophiloid larvae. – 24: *Ochthebius,* dorsal view (left mandible removed); 25: *Ochthebius,* ventral view (mandibles and antennae removed); 26: mandible of *Ochthebius;* 27: *Helophorus,* dorsal view; 28: *Helophorus,* ventral view (mandibles and antennae removed); 29: nasale of *Laccobius;* 30: nasale of *Hydrobius.* (Figs 24-26 redrawn after Böving & Henriksen, 1938; figs 27 & 28 after Angus, 1973). For abbreviations: see p. 23.

20

Maxillae with the usual 3 proximal segments: cardo, stipes and palpiger, bearing the 3-segmented maxillary palpi. In Hydraenidae the stipes and palpiger are fused to a single plate, as are their two appendages (lacinia and galea, respectively) proximally. In the other families the stipes and palpiger are well separated, the latter being almost cylindrical, resembling the palpal segments. S.percheidae have distinct lacinia and galea; Hydrochidae a rudimentary lacinia and no galea; Hydrophilidae and Georissidae are without lacinia, but normally with a small galea.

Labium with 2-segmented labial palpi. Mentum and submentum separate, but in some genera fused (e.g. *Hydrochus, Anacaena*). In some groups (Hydraenidae, Spercheidae, Hydrochidae) with a large separate sclerite (gula) posterior to submentum, and laterally delimited by two gular sutures. In Georissidae and Hydrophilidae the gular sutures have fused to a single median suture, and there is no gular sclerite.

In Hydraenidae (Fig. 31) and Hydrochidae (Fig. 35) each body segment is covered by a soft smooth dorsal and ventral plate. In other groups these plates are reduced, or at least divided into a number of smaller sclerites (except the prothoracal sclerite, which is almost always rather complete). The cuticle (between the sclerites) is membranous, or often coriaceous, asperate and folded, with short felt-like pubescence. The segments (at least on abdomen) often have lateral outgrowths, that may be verruciform, or (*Berosus*) even prolonged into long filaments serving as gills.

The number of abdominal segments is 10. In Hydraenidae the 10th segment forms a distinct, ventrally directed pygopod. In Spercheidae (Fig. 32) it is strongly reduced and very indistinct. In most Hydrophilidae both 9th and 10th abdominal segments are small, to a great extent concealed by the 8th segment; segments 8-10 (or 8 and 9 alone) forming a terminal cavity (breathing pocket), into which the spiracles of 8th segment open. 9th segment with a pair of 1-3-segmented urogomphi. These are strongly developed in *Helophorus* (Fig. 34), but are otherwise generally small, and in many species (including most Hydrophilidae) rather indistinct, sometimes (*Spercheus*) missing. The presence of urogomphi is characteristic of the Staphyliniformia, but they are otherwise only found within Adephaga.

The larvae of Hydraenidae are holopneustic, i.e. they have 9 pairs of spiracles (on prothorax and 1st-8th abdominal segments); the spiracles are annular. Larvae of *Helophorus* and *Georissus* are also holopneustic, but the spiracles are biforous (at least in *Helophorus*). The remainder of the larval forms are metapneustic, i.e. they have only one pair of functional spiracles (on 8th segment). In *Berosus* the last pair of spiracles is also rudimentary, so there are no functional spiracles (apneustic type).

The legs are of the normal polyphagan type, i.e. there are only two segments distal to femur (the terminal segment claw-like). In some genera the legs are reduced in size, sometimes (e.g. most Sphaeridiinae) rudimentary or even missing.

Though larvae of almost all North and Central European genera have been described, much work remains to be done before the larvae of all Fennoscandian and Danish species can be determined. Thus, descriptions of the larvae are not included in the systematic part of this work. Immature stages of Hydrophiloidea are thoroughly dealt with in Böving & Henriksen (1938). A provisional key to the British *Helophorus*-larvae (including many Scandinavian species) is given by Angus (1973b).

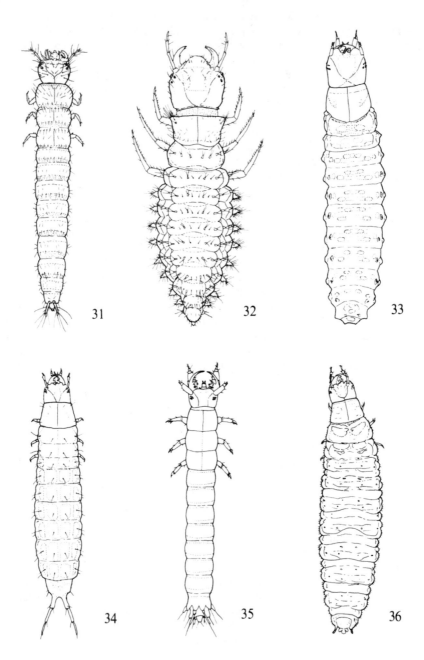

31   32   33   34   35   36

22

List of abbreviations used in the figures

ane – antennal excavation
ant – antenna
bsp – basal piece of aedeagus
car – cardo
clu – antennal club
cly – clypeus
ccv – coxal cavity
cub – cubitus
cup – cupule of antenna
elb – epistomal lobe
ely – elytra
epl – epipleura of elytra
epm1 – epimeron of mesosternum
epm2 – epimeron of metasternum
eye – compound eye
fcf – fronto-clypeal furrow (suture)
fdp – frontal depression
fem – femur
frf – frontal furrow (suture)
fro – frons
gal – galea
gul – gula
hum – humerus
lac – lacinia
lbp – labial palp
lbr – labrum
max – maxilla
med – media
mnd – mandible
mnp – main piece of aegeadus
mnt – mentum
mol – molar area of mandibles
mse – mes-episternum

mss – mesosternum
msx – meso-coxa
mte – met-episternum
mts – metasternum
mtx – meta-coxa
mxp – maxillary palp
nas – nasale
nps – notopleural suture
oce – ocellus
pal – palpiger
ped – pedicel of antenna
pen – penis
pep – pseudepipleura of elytra
prm – paramere
pro – prostheca
prs – prosternum
prt – pronotum
prx – pro-coxa
sca – scape of antenna
scu – scutellum
smt – submentum
sti – stipes
sty – stylus
sut – suture of elytra
tib – tibia
tmb – terminal lobe in aedeagus
tmp – tempora
trc – trochanter
trs – tarsus
1.int – 1st elytral interval
1.str – 1st elytral stria
I-V – abdominal sternites

Figs 31-36. Larvae of hydrophiloids. – 31: *Ochthebius;* 32: *Spercheus;* 33: *Georissus;* 34: *Helophorus;* 35: *Hydrochus;* 36: *Hydrobius.* (Figs 31, 32, 35 & 36 redrawn after Böving & Henriksen, 1938; fig. 33 after van Emden, 1956).

23

Classification and Characteristics

The superfamily Hydrophiloidea (earlier often called Palpicornia) belongs to the largest of the Coleoptera suborders, Polyphaga, and there is general agreement that it should be placed within the Staphyliniformia, which is often divided into 3 superfamilies, viz. – besides Hydrophiloidea – Staphylinoidea and Histeroidea.

Hydrophiloidea (as treated here) may be divided into two well defined groups, both of which seems to be clearly monophyletic: 1) a Hydraenid lineage (including Hydraenidae) characterized by numerous adult synapomorphies, eg. antennal and aedeagal characteristics (cf. description) and reduction of hind wing venation; 2) a Hydrophilid lineage (including Spercheidae, Hydrochidae, Georissidae and Hydrophilidae) which also share numerous adult synapomorphies, of which the most characteristic perhaps is the modification of the antennae (which are basically very uniform, though somewhat aberrant in *Spercheus*) (cf. description).

Due to the marked differences in some characters (also larval) between these two lineages, *and* the similarity between the Hydraenidae and certain staphylinoid beetles regarding some of the characters, the monophyly of Hydrophiloidea (as treated here) has been questioned. Some authors (see eg. Lawrence & Newton, 1982) include the Hydraenidae in the Staphylinoidea, based on e.g. similar reduction of hind wing venation, and some larval similarities (particularly shared with Ptiliidae), which they regard as synapomorphies.

The similarity in larval habitus between the Hydrophilid lineage and Histeroidea has been interpreted as synamorphies by some authors (see eg. Lawrence & Newton, 1982), who consequently included the Histeroid families in the Hydrophiloidea (but excluding the Hydraenidae).

However, the phylogenetic relations within the Staphyliniformia are not evident, and other characters seem to be in strong conflict with this arrangement.

The Hydraenid and the Hydrophilid lineages share many common adult characters, that may very likely be synapomorphies. The aquatic habits (unique within the Staphyliniformia) may indicate monophyly of the Hydrophiloidea as treated here (only a few, clearly derived groups are terrestrial). Connected with this habit are other apomorphies (probably synapomorphies), such as development of dense hydrofuge pubescence on the ventral face of the body (unique within Staphyliniformia, but found in some other aquatic Coleoptera), silk glands for construction of egg cocoons (Lawrence & Newton, 1982), and a quite unique development of the antennae (used in respiration, see under biology).

The differences between the antennae of the Hydraenidae (Fig. 96) and the Hydrophilid lineage (Fig. 304) are superficial, and both groups may easily be derived from a common ancestral form with 11-segmented antennae, having 1st and 2nd segments enlarged, 3rd-5th segments small and uniform, 6th segment forming a cupule, and bearing a 5-segmented pubescent club. Actually, some exotic (presumably primitive) Hydraenids have been described that possess such 11-segmented antennae: but in most Hydraenid genera the small segments between 2nd segment (the pedicel) and the cupule have been reduced in number, so usually only one has remained.

In the Hydrophilid lineage the primary event was a reduction of the number of segments in the antennal club to 3; in some genera the number of small segments between the pedicel and the cupule has subsequently also become reduced.

Other obvious synapomorphies are the strong development of the maxillary palpi, which is unique within the Coleoptera (cf. description), and the peculiar division of the epipleural portion of the elytra (cf. description).

In regard to the body shape and general appearance the Hydrophiloidea form a very variable group. It was earlier treated as a single family: Hydrophilidae (s.lat.), but in more recent time a division into more families has taken place: Hydraenidae, Spercheidae, Hydrochidae, Georissidae and Hydrophilidae (s. str.). Sometimes the latter is divided into more families. The inclusion of *Helophorus* and *Hydrochus* within the Hydraenidae (e.g. Lohse, 1971) is artificial, based merely on plesiomorphic characters; as it will appear from the above mentioned Hydraenidae may be the sister-group of the rest of the hydrophiloid families.

Of these, Spercheidae no doubt forms a monophyletic group defined by many adult synapomorphies: expansion of the clypeal margins (the anterior margin being emarginate, but completely concealing labrum in dorsal view), reduction of antennal segments to 7 (apparently 6) in which also the pedicel and the cupule are pubescent; in larvae the urogomphi have disappeared, and the 10th abdominal segment is rudimentary, indistinct. Possibly Spercheidae is the sister-group to all the following families, which may be characterized by a complete fusion of labrum and clypeus in the larvae (in *Spercheus* only partially fused).

Hydrochidae forms a group of quite uniform species, sharing many adult synapomorphies: the characteristic pattern of shallow impressions on the pronotum, the velvety appearance of the ventral face of the body, the semi-transparent lobe attached to the 5th visible abdominal sternite, and the posteriorly markedly closed procoxal cavities. The larval lacinia is rudimentary (but detectable), and the galea missing. This family probably forms a sister-group to the following two families.

Hydrophilidae and Georissidae may primarily be defined by some larval synapomorphies: complete reduction of the lacinia (but with distinct galea), fusion of the gular sutures to a single median suture, division or complete reduction of the abdominal sclerites.

It should be emphasized that the maintenance of Georissidae as a distinct family (as treated here) must be considered tentative. Perhaps it should be placed in Hydrophilidae (near *Helophorus*), or – if one prefers to maintain the family rank – *Helophorus* should be excluded from the Hydrophilidae. A closer study of these groups remains to be carried out, before their relations can be explained satisfactorily.

Georissidae has earlier been placed near the Elmidae but after the *Georissus*-larva was described, there remains not much doubt that it must be included within the Hydrophiloidea (Emden, 1956).

The phylogenetic relations within Hydrophilidae (s.str.) are not obvious, but the traditional division into subfamilies (which I have followed) generally seems to reflect these as monophyletic groups.

Biology

In most species of Hydrophiloidea the eggs are laid in May-June, but in many the egg-laying period is rather long and may last throughout the summer. Some species may lay their egg in the autumn, and the eggs hibernate (otherwise the larvae usually emerge about a week after the eggs are laid).

A few forms (particularly the Hydraenidae) lay their eggs singly, on stones or algae, in or out of the water. The egg may be naked, but is normally covered by a silk net, which may be either open-meshed (some *Ochthebius*), or densely woven (e.g. *Hydraena*). Other forms (e.g. *Hydrochus*) completely enclose the egg in a capsule. In most of the Hydrophiloids, the eggs are laid in groups of 2 or more (even up to 70 in *Hydrophilus*), enclosed in a densely woven (or seldom loose) silk cocoon.

The cocoon is often drawn out into a variably long band or vertical mast (Figs 39-42), that plays an important role in supplying oxygen to the developing eggs, at least in some species (Angus, 1973b).

Most Hydrophilidae place their egg cocoons on shallow water (just above the water surface) attached to vegetation, or in moist ground near the edge of the water. In *Hydrophilus* and *Hydrochara* the cocoons float freely on the water, being not attached to the vegetation. The terrestrial forms within the Sphaeridiinae place their (loosely spun) cocoons in different kinds of decaying organic matter. In a few forms (*Spercheus, Helochares*) the females do not produce such egg cocoons, but instead carry the eggs in a bag on the ventral face of abdomen.

The larvae emerge predominantly in late spring or early summer and generally develop very rapidly, being full grown after about 10-20 days (and two moults).

While the Hydraenid larvae feed on algae, almost all the other larval forms are voracious predators, probably feeding mainly on snails, worms, insect larvae etc. Many larvae show a great extent of cannibalism (at least when they are kept in captivity). The larvae of *Helophorus* subgenus *Empleurus* are phytophagous, and more long-lived; some of these have been recorded as pests of cultivated plants (turnips, wheat, rye).

The larvae of the more primitive forms (e.g. Hydraenidae, *Helophorus*) are terrestrial, or rather, semiaquatic, confined to humid or wet habitats, near the edge of water. Most hydrophilid larvae are aquatic, living in the water, rather than on moist ground near it; these larvae have adapted to the aquatic environment by developing a terminal abdominal breathing pocket, into which the only functional pair of spiracles open (cf. description); in a few forms (*Berosus*) there is no breathing pocket, and the respiration takes place by means of tracheal gills.

Only a smaller group of larvae, most of which belong to the Sphaeridiinae, are terrestrial in the sense that they often live far from water, in various kinds of decaying organic matter, where they feed on small invertebrates, perhaps dipterous larvae, in particular. Possibly more of the sphaeridiine larvae also ought to be considered semiaquatic, as the substrata in which they live are generally quite humid.

Pupation normally takes place in moist ground near the edge of the water, but occasionally pupae of some forms (*Enochrus*) may be found in hollow stems of water plants (Kryger & Sønderup, 1940); the larvae of Sphaeridiinae pupate in the substratum in which they live, or in the earth beneath it. The pupal stage is normally short, often lasting for no more than a week.

The adults emerge in the summer. Contrary to the larvae, they feed almost exclusively on plants or decaying plant debris. Hibernation almost always takes place in the adult stage (though also the eggs may sometimes hibernate, as mentioned above). The Sphaeridiinae, or at least some of them, are unusual within the Hydrophiloidea, in that they have two generations per year (one is the norm), larvae emerging in spring giving rise to a summer generation; hibernation still takes place in the adult stage.

As with their larvae, the adults of most species are aquatic or semiaquatic; the Sphaeridiinae live (also as adults) in decaying organic matter, a few forms being confined to seashores. Also within *Helophorus* are found a few terrestrial forms (subgen. *Empleurus*).

Generally, the aquatic Hydrophiloidea live in shallow or very shallow water. Apart from a few forms (e.g. *Hydrophilus, Berosus*) they are rather poor swimmers, more adapted to climbing among the vegetation. Some torrenticolous species (mainly *Hydraena* spp.) are often found under stones or the like in the water.

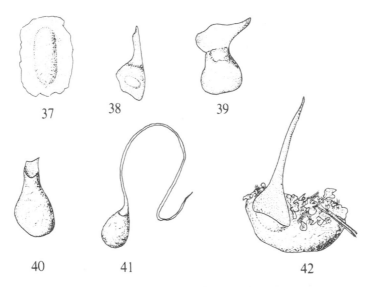

Figs 37-42. Egg cocoons of hydrophiloids. – 37: *Hydraena* sp.; 38: *Hydrochus* sp.; 39: *Hydrobius fuscipes;* 40: *Helophorus* sp.; 41: *Helophorus obscurus;* 42: *Hydrophilus aterrimus*. (Figs 37-40 redrawn after Böving & Henriksen, 1938).

27

The adults of the aquatic forms show a mode of respiration that is unique within the Coleoptera. When the beetle is in the water it is able to maintain a ventral air bubble, due to the dense hydrofuge pubescence covering the ventral surface of the body (see description); the bubble is delimited laterally by the somewhat overhanging epipleural portion of the elytra. In order to renew its supply of oxygen the beetle ascends to the water surface, where it »hangs« in an oblique position, with the head immediately below the surface, and the abdomen hanging slightly down. When sitting in this position it directs the antennae (or often only one of them) towards the dorsal face, but with the club bending ventrally and lying closely up against the tempora. The hydrofuge pubescence of the antennal club and of the tempora makes it possible for the ventral air bubble to continue into the small »tube« between these, so – when the water surface film is broken by the antenna – contact is made between the ventral air bubble and the atmospheric air.

When this contact is established, the beetle – still hanging in the same position – starts to pump new air into the tracheae. When the air is sufficiently renewed, the beetle dives again. Apparently no mechanical respiration takes place when the beetle is not in contact with the atmospheric air.

The way in which the surface film is broken is subject to some variations (Orchymont, 1933a), but the principle of respiration is the same within the different groups of Hydrophiloidea.

Many species, particularly the smaller species, may – when they are disturbed – ascend to the water surface, where they may be turned upside down due to the ventral air bubble; in this position they are able to climb across the under side of the surface film.

Most of the Hydrophiloids are active flyers, being particularly active in May-June; they fly predominantly towards evening, often before sunset, but there are many species that readily fly during most of the day. Some species apparently do not fly (though wings seem to be fully developed). A few species are brachypterous, or even apterous (cf. description).

Both males and females of many species are able to produce sounds by friction between a stridulatory area on the ventral face of the elytra (cf. description) and the lateral portion of the 1st complete abdominal segment. Stridulation may be initiated when the beetle is stressed, but apparently it is more important in association with reproductive behavior. In some genera, pre-mating stridulatory sounds have been shown to be specifically distinctive, and they may play a role in preventing interbreeding between closely related species (Ryker, 1972).

Zoogeography

On a world-wide scale the Hydrophiloidea comprise almost 3500 described species. Each of the families are known from all the major zoogeographical regions of the world. Short notes on the distribution of each genus are given after the descriptions. Moving from the temperate zones towards the warmer climates many genera show a pronounced increase in the number of species, and the hydrophiloids no doubt form a group where large numbers of species are yet to be described.

At present, a total of 118 species is known from Fennoscandia and Denmark. By far the most of these have a wide palearctic (or at least European) range, and have their northern limits in Central or North Scandinavia.

Some species showing a Mediterranean (e.g. *Limnoxenus niger*) or West European (e.g. *Helochares punctatus*) type of distribution are found only in southernmost Scandinavia.

A smaller group of species exhibits a northern pattern of distribution. Some of these (e.g. *Helophorus glacialis*) are boreo-montane, confined to North Fennoscandia and the mountainous regions of Central Europe, while others (e.g. *Helophorus pallidus*) are boreal species with a wide palearctic (predominantly eastern) distribution.

Several species (e.g. *Ochthebius marinus, Helophorus tuberculatus, Cercyon marinus*) are holarctic; some species of *Sphaeridium* and *Cercyon* also show a holarctic type of distribution, but their occurrence outside the Palearctic region is due to introduction by man (Smetana, 1978).

A few species are particularly interesting, as they have not been recorded outside North (or Central) Fennoscandia, viz., *Ochthebius stockmanni* B.-Br., *O. nilssoni* Hebauer, *Helophorus strandi* Angus, *Cercyon emarginatus* Baranowski, *C. borealis* Baranowski and *Hydrobius arcticus* Kuw. Another species, *Ochthebius kaninensis* Popp., is known only from North-east Finland and the Kanin Peninsula. It is possible that some of them may be endemic to Fennoscandia, but more likely the absence of records outside this area is due to the fact that the descriptions of these species (except *Hydrobius arcticus* and *Ochthebius kaninensis,* the latter having hitherto been unrecognized) are comparatively (or even very) recent. Probably they will be shown to be widespread throughout the north of the East Palearctic region.

Nomenclature

The nomenclature presented here largely agrees with that in the recent catalogue of the Fennoscandian and Danish Coleoptera (Silfverberg, 1979). All generic names are in accordance with those presented there, and only in a few cases have I found it necessary to change a specific name.

Re-examination (including study of genital characters) of the old or very old type material has already been carried out by previous authors, and by now the nomenclature in Hydrophiloidea seems to be rather well established, at least on the specific level. Thus I have had to examine very few types, and they have merely confirmed the current use of the respective names.

The generic nomenclature also seems to be quite stable, but the type-species of some genera have been a source of some confusion, and study of older literature has revealed that in a few genera, or rather subgenera, there appeared to be no valid type designation. In these cases type-species are designated in the present work.

It should be mentioned that a few of the generic names used in the systematic part (and the subgeneric names, in particular), are not strictly valid according to the International Code of Zoological Nomenclature. These are: *Ochthebius* subgenus *Homalochthebius* (misidentified type-species), *Helophorus* (more problems, see after the generic description), *Enochrus* (s.str.) (misidentified type-species), *Enochrus* subgenus *Methydrus* (misidentified type-species). However, these are names that have been currently used for more than the last 50 years (or even 100 years), and are generally accepted by recent authors. Thus, instead of making some (perhaps) unnecessary changes to these names (which may only cause confusion), I have preferred to follow the general interpretation. These cases ought to be referred to the International Commission of Zoological Nomenclature.

In the present work, junior synonyms (on both generic and specific level) are included only in those cases, where they have been used in Scandinavian literature, or if for other reasons they have been considered of special interest.

Identification

Most of the hydrophiloids are small or rather small species. Many are easily recognized, but in some genera a close examination of characters requiring magnifications of about 50-100 x is necessary. Among the species with furrows or impressions on head and pronotum, specimens are often found, that are covered with a clayey or muddy substance. Such individuals must be cleaned carefully (often a drop of water is enough to dissolve the substance) as this muddy coat often conceals just the characters that are most important for determining the species (e.g. sculpture of head and pronotum).

In some genera (e.g. *Hydraena, Helophorus, Laccobius*) identifications of the species may be difficult, so examination of the male genitalia can be necessary, or at least very helpful. A few remarks on the preparation should be made.

In fresh or softened material the aedeagus is rather easily extracted from the abdomen by means of a sharp insect pin. The extracted genitalia is placed in water and is then cleaned from adhering sclerites, ligaments and muscles. As the genitalia are generally small and not very strongly sclerotized it is not recommended to make dry preparations of them (as is often done within the Coleoptera); rather they ought to be placed in a drop of mountant, such as euparal; this mountant is advantageous in that it clears the aedeagus so the internal structures can be examined. But euparal is soluble in alcohol so before it is mounted the aedeagus should be transferred into alcohol for a few minutes. The genital-figures given in the present work are based on genitalia prepared in this way, and mounted so that they are seen in dorsal or (most Hydraenidae) lateral view.

The diagnostic characters given in the keys of genera and higher taxa are only of value for the North European fauna.

Acknowledgements

During my work on this volume I have received much support and assistance. For the kind loan of material from various institutions and for faunistic information I am most grateful to Mr. O. Lomholdt, Zoologisk Museum, Copenhagen; Dr. P. Gjelstrup, Naturhistorisk Museum, Århus; Dr. N. Haarløv, Den Kgl. Veterinær- og Landbohøjskole, Copenhagen; Dr. R. Danielson, Zoologiska Museet, Lund; Dr. G. Andersson, Naturhistoriska Museet, Göteborg; Dr. S. Jonsson, Dept. of Ent., University of Uppsala, Uppsala; Dr. K. Sund, Zoologisk Museum, Oslo; Dr. L. Greve Jensen, Zoologisk Museum, Bergen; Dr. A. Fjellberg, Tromsø Museum, Tromsø; and to Dr. O. Biström and Dr. H. Silfverberg, Universitetets Zoologiska Museum, Helsingfors.

Further, I am most indebted to the following persons for the loan or donation of material from their private collections and for faunistic or biological information. Denmark: Mr. M. Holmen, Copenhagen; Mr. S. Kristensen, Them; Mr. V. Mahler, Århus; Mr. G. Pritzl, Strøby; Mr. O. Vagtholm, Billund. Sweden: Mr. R. Baranowski, Lund; Mr. J. R. Bergvall, Gällö; Mr. B. Ehnström, Uppsala; Mr. B. Ericson, Höör; Mr. B. Henriksson, Edsbyn; Mr. C. Holmqvist, Stockholm; Mr. T.-E. Leiler, Vallentuna; Mr. Å. Lindelöw, Storvreta; Mr. S. Lundberg, Luleå; Mr. A. Nilsson, Vansbro; Dr. A. Nilsson, Umeå; Mr. S. E. Nilsson, Arvidsjaur; Mr. T. Palm, Malmö; Mr. B. Persson, Lycksele; Mr. S. Persson, Landskrona; Mr. M. Sörensson, Lund; Mr. B. Weidow, Skara. Norway: Mr. O. Hanssen, Dragvoll. Finland: Mr. T. Clayhills, Pargas; Mr. T. Ilvessalo, Turku; Mr. J. Muona, Oulu; late Dr. C. von Numers, Grankulla; Mr. M. Pohjola, Vuorentausta; Mr. I. Rutanen, Hyvinkää. England: Dr. R. B. Angus, Alderhurst. Germany: Mr. F. Hebauer, Deggendorf; Dr. G. A. Lohse, Hamburg; Mr. H. Meybohm, Stelle.

I also wish to thank Dr. A. Smetana, Ottawa, Canada for information concerning some North American spcies.

Special thanks are due to Mr. G. Pritzl, Strøby for valuable and inspiring discussions and critical remarks during the work, to Dr. M. Luff, Newcastle for linguistic correction of the manuscript, and to the managing editor, Dr. L. Lyneborg, for much support.

Key to families of Hydrophiloidea

1 Antennae 9-segmented, with 5 distal segments pubescent, forming a loose club (Fig. 44). Abdomen with 6 or 7 visible sternites. Small species, 1.0-2.8 mm . . . . Hydraenidae (p. 32)

– Antennae with only 3 distal segments pubescent, forming a loose or compact club (eg. Figs 206, 209), or apparently 6-segmented with only basal segment glabrous (Fig. 117). Abdomen almost always with only 5 visible sternites 2

2 (1) Antennae (Fig. 117) apparently 6-segmented (in fact 7-segmented with most indistinct 3rd segment), with 5 distal

Family Hydraenidae

Head sometimes with two well developed ocelli near inner margin of eyes. Scutellum distinct. Prosternum well developed, not concealed by procoxae, on each side with a variable antennal excavation. Procoxal cavities closed or open posteriorly. Abdomen with 6 or 7 visible sternites. Antennae 9- to 11-segmented, with 5 distal segments forming a densely pubescent club (in a few aberrant exotic forms with only 2 segments in the club). Tarsi 5-segmented with basal segment minute, often very indistinct; 2nd and 3rd segments may often also be rather small, mutually very closely articulating, so that tarsi appear 3-segmented. Claw segment at least about as long as the preceeding segments (except in *Limnebius*). All trochanters distinct. Venation of hind wings reduced, of the staphylinoid type (Fig. 23). Aedeagus (Fig. 14) basically consisting of a well sclerotized tubular main piece terminating in a mobile lobe, and bearing 2 lateral parameres; parameres in some forms rudimentary or even missing, in others to some extent fused with the main piece, forming a very complex structure.

The family has often been divided into 2 subfamilies: Hydraeninae (including *Hydraena, Ochthebius* and some other (exotic) genera) and Limnebiinae (including only *Limnebius*). Perkins (1980) points out that this division is most likely artificial, based on plesiomorphic characters. He gives some characters (probably synapo-morphic), that indicate a closer relationship between *Hydraena* and *Limnebius*. Con-

32

sequently, he includes *Limnebius* in Hydraeninae, and places *Ochthebius* (and a few similar (exotic) genera) in a separate subfamily: Ochthebiinae.

A rather large family, comprising almost 1000 described species, and known from all parts of the world. 3 genera occur in North and Central Europe.

Key to genera of Hydraenidae

1   Body contour evenly curved, pronotum not narrowed po-
    steriorly, its surface even, without impressions .... *Limnebius* Leach (p. 64)
–   Body contour interrupted between pronotum and elytra,
    pronotum narrowed posteriorly, its surface rather uneven,
    with distinct impressions 2
2 (1) Maxillary palpi rather short, their terminal segment mi-
    nute, much smaller than penultimate (Fig. 45). Pronotum
    bordered by a narrow hyaline membrane .... *Ochthebius* Leach (p. 33)
–   Maxillary palpi very long and slender, their terminal seg-
    ment well developed, longer than penultimate (Figs 71-75).
    Pronotum without marginal hyaline membrane . *Hydraena* Kugelann (p. 50)

Genus *Ochthebius* Leach, 1815

Ochthebius Leach, 1815, *in* Brewst. Edinb. Enc. 9: 95.
    Type species: *Elophorus marinus* Paykull, 1798, by subsequent designation (d'Or-chymont, 1942b).

Body contour interrupted between pronotum and elytra. Head with large protruding eyes, and a transverse furrow separating clypeus and frons; the latter on each side with an interocular pit-like depression, and a short longitudinal depression or furrow postero-medially. Sometimes with two well developed ocelli near inner margin of eyes. Anterior margin of labrum truncate, sometimes with a small median emargination.

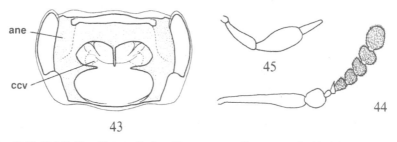

Figs 43-45. *Ochthebius dilatatus* Stph. – 43: prosternum (legs removed); 44: right antenna; 45: maxillary palpus.

33

Pronotum bordered by a narrow hyaline membrane; middle pronotal portion raised, delimited laterally by a variably deep longitudinal depression, and provided with a characteristic pattern of depressions, varying from one subgenus to another (Figs 46-53). Lateral margins of pronotum posteriorly with a more or less pronounced excision, which is filled up (or at least bordered) by the marginal membrane. Elytra in all the species treated here (except *evanescens*) with 10 distinct striae or series of punctures; in some species additionally with a short intercalary row of punctures at base, between 1st and 2nd stria. Lateral margins of elytra forming a sharp ridge, which in most species disappears posteriorly, but (subgenus *Enicocerus*) may be distinct to apex. Ventral surface of body dull, pronouncedly shagreened and with dense hydrofuge pubescence; metasternum sometimes with a small glabrous and shining field on middle portion. Procoxal cavities open posteriorly (Fig. 43). Abdomen with 7 visible sternites; the last two sternites more shining and more sparsely pubescent than preceeding sternites. Maxillary palpi (Fig. 45) rather short, terminal segment minute, much smaller than penultimate. Antennae (Fig. 44) 9-segmented with 5-segmented pubescent club. Legs more or less slender, in some species very long.

A large genus with a world wide distribution, comprising about 300 known species. More than 25 species occur in Central and North Europe.

Key to species of *Ochthebius*

1	Pronotum rather wide, about 2/3 as wide as long, with a very pronounced latero-basal excision, very abruptly and strongly narrowed posteriorly (Figs 46, 47, 52, 53)	2
–	Pronotum narrower, with a much narrower and less abrupt latero-basal excision (Figs 48-51)	8
2 (1)	Pronotum with two strongly depressed furrows, one before and one behind middle, connected by a very sharp and narrow median longitudinal furrow (Figs 52, 53). Very small species, 1.0-1.3 mm	3
–	Pronotum without transverse furrows, but with a long conspicuous median longitudinal furrow, on each side of which are two pit-like depressions, one behind the other (Figs 46, 47). Larger species, 1.6-2.1 mm	4
3 (2)	Lateral margins of pronotum (anterior to latero-basal excision) evenly rounded (Fig. 52) . . . . *exaratus* Mulsant
–	Lateral margins of pronotum (anterior to latero-basal excision) in middle distinctly angularly emarginate (Fig. 53) . . . . *narentinus* Reitter
4 (2)	Elytra more than 1/3 longer than wide (Figs 54, 55), series of punctures moderately strong. Intercalary stria usually rather long, consisting of normally 5-8 punctures	5
–	Elytra shorter and more convex, less than 1/3 longer

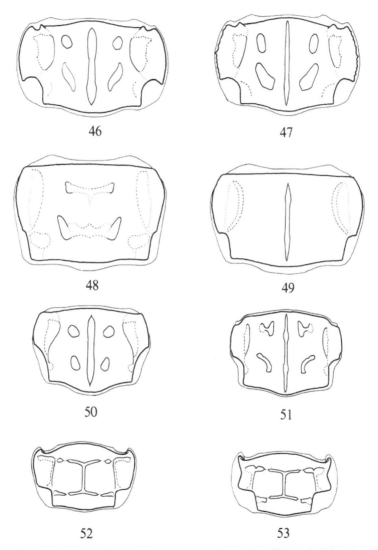

Figs 46-53. Pronotum of *Ochthebius.* – 46: *bicolon* Germ.; 47: *stockmanni* B.-Br.; 48: *marinus* (Payk.); 49: *minimus* (F.); 50: *nilssoni* Hebauer; 51: *gibbosus* Germ.; 52: *exaratus* Muls.; 53: *narentinus* Reitt.

than wide. Intercalary striae shorter or missing, consisting of no more than 4 punctures 6

5 (4) Pronotum almost as wide as elytra (Fig. 55). Anterior margin of labrum distinctly emarginate .... 1. *auriculatus* Rey
– Pronotum narrower, elytra at least 1/5 wider than it (Fig. 54). Anterior margin of labrum truncate, without distinct emargination .... 2. *dilatatus* Stephens

6 (4) Median pronotal furrow very narrow (Fig. 47). Elytra without intercalary striae (sometimes with a single basal puncture) 7
– Median pronotal furrow wider (Fig. 46). Elytra with a short intercalary stria consisting of 3-4 punctures ..... 3. *bicolon* Germar

7 (6) Pronotum shining, with fine or very fine, rather sparse punctuation .... 5. *kaninensis* Poppius
– Pronotum with dense and strong punctuation ....
.... 4. *stockmanni* Balfour-Browne

8 (1) Pronotum with a long conspicuous median longitudinal furrow, without large transverse impressions (Figs 49-51)... 9
– Pronotum without conspicuous median furrow, but

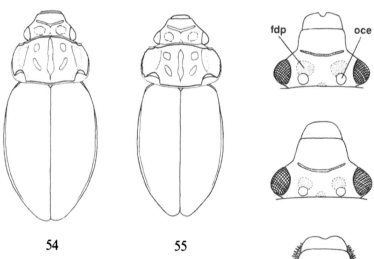

54   55

Figs 54, 55. Outline of *Ochthebius*. – 54: *dilatatus* Stph.; 55: *auriculatus* Rey.
Figs 56-58. Head of *Ochthebius*. – 56: *pusillus* Stph.; 57: *viridis* Peyr.; 58: *bicolon* Germ.

with two large transverse impressions, one before and
one behind middle (Fig. 48) 11

9 (8)   Raised middle portion of pronotum on each side of median furrow with two conspicuous pit-like depressions,
one behind the other (Figs 50, 51) 10

–   Raised middle portion of pronotum without such depressions (Fig. 49) .... 6. *minimus* (Fabricius)

10 (9)   Elytra wide and short, about 1/2 wider than pronotum, strongly punctato-striate .... *gibbosus* Germar

–   Elytra rather elongate, only about 1/4 wider than pronotum, finely punctato-striate .... 7. *nilssoni* Hebauer

11 (8)   Elytra punctuated only in anterior half, the punctures
very fine and sparse, somewhat irregular, hardly form-
ing distinct series .... 11. *evanescens* J. Sahlberg

–   Elytra punctato-striate, the punctures much stronger,
distinct to apex 12

12 (11)   Small black species, 1.4-1.7 mm................................. 13

–   Larger species, 1.8-2.1 mm, with brownish, seldom almost black elytra 14

13 (12)   Anterior margin of labrum evenly rounded, not emarginate (Fig. 57) .... 10. *viridis* Peyron

–   Anterior margin of clypeus with a small, but very distinct median emargination (Fig. 56) .... *pusillus* Stephens

14 (12)   Legs fairly long and slender (Fig. 64). Elytra brown or
yellowish brown, head and pronotum often rather
bright metallic. ♂: aedeagus as in Fig. 66 .... 8. *marinus* (Paykull)

–   Legs, especially tarsi, shorter (Fig. 63). Elytra dark or
almost blackish brown. Metallic reflections of head and
pronotum rather dull. ♂: aedeagus as in Fig. 65 ..... 9. *lenensis* Poppius

Subgenus *Enicocerus* Stephens, 1829

Enicocerus Stephens, 1829, Ill. Brit. Ent. Mandib. 2: 196.
   Type species: *Enicocerus viridiaeneus* Stephens, 1829 (= *Ochthebius exsculptus*
   Germar, 1824), by monotypy.
Henicocerus Stephens; Bedel 1881, Faune Col. bass. Seine 1: 291 (emend.).

Lateral margins of pronotum excised in about posterior half, but less abruptly than in
Asiobates; the excision narrowly bordered by the marginal membrane. Pronotum (Fig.
51) with a conspicuous median longitudinal furrow, which may be reduced posteriorly,
and on each side of this with an additional depression, or (as in the species treated
here) with a pattern of depressions very similar to that of *Asiobates.* Marginal elytral
ridge distinct to apex. 2nd antennal segment short, enlarged distally (cup-shaped).
Terminal segment of maxillary palpi very small, shorter than in the other subgenera.

Ochthebius gibbosus Germar, 1824
   Fig. 51.

Ochthebius gibbosus Germar, 1824, Ins. spec. nov.: 93.

1.2-1.4 mm. Black, without distinct metallic hue. Head shining, impunctate. Anterior margin of labrum weakly emarginate. Pronotum comparatively small, strongly convex, with extremely fine or indistinct punctuation, at most near posterior angles and at anterior margin with fine rugulose punctuation; median longitudinal furrow sharp and narrow, reaching almost from anterior to posterior margin, on each side with two small pit-like depressions, one before and one behind middle. Elytra very short and wide, about 1/2 wider than pronotum, strongly convex and strongly punctato-striate, interstices rather convex; intercalary striae absent. Appendages yellowish red to reddish brown; antennal club (and often also maxillary palpi) slightly darker.

Distribution. So far not recorded from Scandinavia. – West and Central European species, from the Pyrenees, North Italy and Hungary (Siebenbürgen) through mountainous regions of Central Europe, north to the Polish and North German lowland (not in Great Britain). Perhaps it may also be found in southern Scandinavia.

Biology. The species lives in streams, under stones or in moss, and on sandy or gravelly banks of streams.

Subgenus *Asiobates* Thomson, 1859

Asiobates Thomson, 1859, Skand. Col. 1: 15.
   Type species: *Ochthebius rufomarginatus* Erichson, 1837 ([emend. of *Ochthebius rufimarginatus* Stephens, 1829] = *Ochthebius bicolon* Germar, 1824), by original designation.

Lateral margins of pronotum pronouncedly excised in about posterior third, very abruptly and strongly narrowed; the excision filled up by the marginal membrane. Pronotum (Figs 46, 47) with conspicuous median longitudinal furrow, reaching almost from anterior to posterior margin; on each side of this with two pit-like depressions, one behind the other. Marginal elytral ridge disappearing posteriorly. 2nd antennal segment short, not enlarged distally.

1. *Ochthebius auriculatus* Rey, 1886
   Fig. 55.

Ochthobius auriculatus Rey, 1886, Annls Soc. linn. Lyon 32 (1885): 45.

1.8-2.1 mm. Reddish to pitchy brown, lateral portions of pronotum paler, reddish. Dorsal surface without metallic hue. Head very finely punctured; anterior margin of labrum distinctly emarginate. Pronotum rather finely punctured, especially on raised middle portion, the punctuation usually slightly denser than in *dilatatus*. Lateral margins of pronotum finely crenulate. Elytra only about 1/10 wider than pronotum, rather strongly punctato-striate, the punctures becoming distinctly finer apically; each puncture with a fine curved seta; interstices almost flat or only weakly convex. Intercalary stria present, consisting of 5-8 punctures. Appendages yellowish red, antennal club a little darker.

♂: Protarsi weakly dilated basally. Outer edge of mandibles with extremely fine and short, hardly visible setae. Elytra shining, only with very weak transverse rugulose microsculpture.

♀: Protarsi and mandibles simple. Elytra a little more dull than in ♂, with more distinct rugulose and reticulate microsculpture.

Distribution. Rare in Scandinavia, only more frequent in SW. Jutland; in Denmark known from SJ, EJ, WJ, NEJ and F; in Sweden only a few localities from Sk., Hall. and Boh.; not in Norway and Finland. – A coastal species, ranging from West France (Bay of Biscay) and the British Isles to southern Scandinavia.

Biology. The species is a true halobiont, that lives exclusively in salt marsh areas, at the edges of shallow brackish pools, most likely in wet mud on the banks, rather than in the water. Apparently it requires a higher salinity than *dilatatus,* and it is usually found in habitats regularly flooded by the tide. Mainly in spring, but also later in summer and autumn, sometimes very abundant.

Note. In the catalogue (Lindroth, 1960) the species was mentioned from Finland (St, Ta), but this is no doubt due to an error. Compared to the known area of distribution, and to the ecology of the species, this record is improbable. The collections of the Helsinki museum do not contain Finnish specimens of *auriculatus,* and apart from Lindroth's records I have not seen it recorded from Finland.

2. *Ochthebius dilatatus* Stephens, 1829
   Figs 43-45, 54, 59.

Ochthebius dilatatus Stephens, 1829, Ill. Brit. Ent. Mandib. 2: 114.
Ochthebius impressicollis Laporte de Castelnau, 1840, Hist. Nat. An. Art. 2: 48.

1.8-2.1 mm. Piceous to black, dorsal surface with a faint bronze hue. Head very finely punctured, the punctures becoming a little larger posteriorly. Anterior margin of labrum truncate, hardly emarginate. Pronotum with rather fine punctuation, especially on raised middle portion, distance between punctures larger than their diameter; median longitudinal furrow rather broad. Lateral margins of pronotum hardly crenulate. Elytra about 1/5 wider than pronotum, rather strongly punctato-striate, punctures becoming finer apically; each puncture with a fine curved seta; interstices almost flat or only very weakly convex; intercalary row of punctures distinct, normally consisting of 5-8 punctures (the number may accidentally be reduced to even a single puncture). Antennae yellow with brown club, palpi brownish, legs yellowish red.

♂: Protarsi weakly dilated basally. Outer edge of mandibles with a distinct tuft of short stout setae (as Fig. 58). Aedeagus, Fig. 59.

♀: Protarsi and mandibles simple.

Distribution. Fairly common along the Danish coasts; in Sweden only known from a few localities in Sk. and Hall.; not in Norway and Finland. – Widespread along the coasts of West and South Europe; from Scandinavia, the Netherlands, France, the British Isles, south to the Mediterranean (main area of distribution), eastwards to Asia Minor and the Caspian Sea.

Biology. A halobiontic species, found exclusively on salt marshes along the coasts. All Scandinavian records are from brackish, mainly stagnant water, typically at the edges of shallow pools, or at the border of drainage canals, among vegetation or in wet mud, usually fairly abundant. In southern Europe the species is more euryoecious, and is also found in fresh water, even in small streams. Occasionally taken in drift on the seashore. March-June, August-October.

3. *Ochthebius bicolon* Germar, 1824
   Figs 46, 58, 60; pl. 1: 5.

Ochthebius bicolon Germar, 1824, Ins. spec. nov.: 92.

1.6-1.8 mm. Piceus to black, lateral portions of pronotum reddish. Dorsal surface with a faint dark bronze hue. Head shining, very finely punctured, between eyes and frontal depressions finely punctured; anterior margin of labrum distinctly emarginate. Pronotum (Fig. 46) denser and a little more strongly punctured than in *dilatatus,* distance between punctures not larger than their diameter; median longitudinal furrow rather broad; lateral margins of pronotum not (or very weakly) crenulate. Elytra shorter, more convex and with more rounded sides than in *dilatatus,* strongly punctato-striate, punctures becoming smaller apically; each puncture with a very fine curved seta; interstices narrower than in *dilatatus,* rather convex. Intercalary stria present, normally consisting of 3 or 4 punctures. Antennae yellow with brown club, palpi brownish, legs yellowish red.

♂: Protarsi weakly dilated basally. Outer edge of mandibles with a distinct tuft of short stout setae (Fig. 58). Elytra shining, with extremely fine and indistinct reticulation. Aedeagus, Fig. 60.

♀: Protarsi and mandibles simple. Elytra rather dull, with very distinct transverse reticulate microsculpture.

Distribution. In Denmark relatively local, but found in most parts; southern Sweden (north to Vstm.); in Norway only in Ø, AK and VE, and a single specimen from the north (Fi: Valjok); in Finland only Al. – Ranges from Fennoscandia and the British Isles to North and Central France and Switzerland, eastwards to Slovakia.

Biology. In rivers or streams, predominantly in the more slowly flowing water at the grassy edges among vegetation, or on the banks, often found in numbers. Only occasionally (and probably accidentally) found in stagnant water. Records of *bicolon* from brackish water are probably due to confusion with *dilatatus,* and it seems to be an exclusive inhabitant of fresh water. Mostly in spring, but also later in summer and autumn.

Note. Lindroth (1960) recorded *bicolon* from Ab and N in Finland. I have examined the specimens standing as *bicolon* in the collection of the Helsinki museum. It appeared that only specimens from Al were this species; specimens from Ab and N proved to be *stockmanni.*

4. *Ochthebius stockmanni* Balfour-Browne, 1948
Figs 47, 61A.

Ochthebius stockmanni Balfour-Browne, 1948, Notul. ent. 28: 95.

1.6-1.8 mm. Very similar to *bicolon.* Black, lateral portions of pronotum often a little paler, brown or sometimes reddish; dorsal surface without metallic hue. Head shining, very finely punctured, between frontal depressions and inner margin of eyes on each side with rugulose punctuation (which is stronger than in *bicolon*); median longitudinal furrow of pronotum narrow and rather sharp (Fig. 47); lateral margins crenulate, pronotal punctuation stronger and denser than in *bicolon.* Elytra without intercalary striae, or at most with a single puncture (only exceptionally two) at base between 1st and 2nd stria; interstices less convex than in *bicolon.* Appendages yellowish red, antennal club (and often maxillary palpi) reddish brown.

♂: Protarsi weakly dilated basally. Outer edge of mandibles without setae. Elytra shining, at most very indistinctly reticulate or rugulose. Aedeagus, Fig. 61A.

♀: Protarsi and mandibles simple. Elytra shining, though a little more dull than in ♂, weakly or very weakly reticulate.

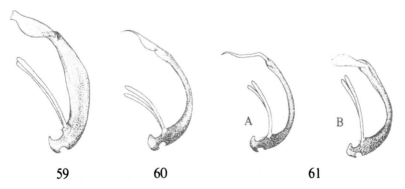

59   60   61

Figs 59-61. Male genitalia of *Ochthebius.* - 59: *dilatatus* Stph.; 60: *bicolon* Germ.; 61A: *stockmanni* B.-Br.; 61B: *kaninensis* Popp.

Distribution. So far only known from a limited area in South Finland (Ab, N). – The species has not been recorded outside Finland.

Biology. At the edges of rivers or streams, perhaps preferring rather slowly running water. The species has also been taken in flight (with car net), and occasionally in drift on the seashore (I. Rutanen, *in litt.*). April, June.

5. *Ochthebius kaninensis* Poppius, 1909, stat. rev.
   Fig. 61B.

Ochthebius kaninensis Poppius, 1909, Acta Soc. Fauna Flora fenn. 31 (8): 29.

1.6-1.9 mm. Similar to *bicolon* and *stockmanni*, but at once recognized from these (as well as all other species of the subgenus) by the very fine pronotal punctuation. Dark reddish brown, dorsal surface without metallic hue. Head shining, very finely punctured, also between inner margin of eyes and frontal depressions; the latter comparatively small, but very deep; anterior margin of labrum emarginate. Pronotum shining, punctuation on raised middle-portion very fine and sparse, not stronger than on head; only near posterior angles and along anterior margin with a few stronger punctures; median longitudinal furrow narrow and sharp; lateral margins crenulate. Elytra without intercalary stria, or at most with a single puncture at base, between 1st and 2nd stria; interstices as in *stockmanni*. Appendages yellowish red, antennal club (and often maxillary palpi) darker.

  ♂: Protarsi weakly dilated basally. Outer edge of mandibles without setae. Elytra shining, at most very indistinctly reticulate or rugulose. Aedeagus, Fig. 61B.

  ♀: Protarsi and mandibles simple. Elytra shining, though a little more dull than in ♂, very weakly reticulate.

Distribution. In our area known only from a single specimen found near Kuusamo in North Finland (Ks), 16. vii. 1968. – Besides this record, the only modern specimens that I know of, are those of the type-series: 5 specimens from the Kanin Peninsula (Kambalniza) in the USSR.

Biology. Apparently a stenothermic, pronouncedly cold-loving species, living in running fresh water, perhaps preferring smaller streams; the type-series was taken at the edge of a tundra-stream ("...im Schlamme am Ufer eines Tundra-Baches..." (Poppius, 1909: 30)). According to Angus (*in litt.*) the species occurs as a pleistocene fossil in England, from about 44.000 years ago – a very cold phase of the last glaciation.

Note. Poppius described *kaninensis* as a good species (though with some reservation). Its specific rank was suppressed by Knisch (1924), who treated it as a subspecies of *bicolon* Germ. Subsequently, there seems to have been taken no notice of *kaninensis* (besides Angus' fossil records), though it is indeed quite characteristic. There is no doubt that it should be considered a valid species.

Subgenus *Aulacochthebius* Kuwert, 1887

Aulacochthebius Kuwert, 1887, Dt. ent. Z. 31: 376.
Type species: *Ochthebius exaratus* Mulsant, 1844, by monotypy.

Lateral margins of pronotum pronouncedly excised in about posterior third, abruptly and strongly narrowed; the excision filled up by the marginal membrane. Pronotum (Figs 52, 53) with two strongly depressed and sharp transverse furrows reaching across whole width (though narrowly interrupted on each side), and with a sharp median longitudinal furrow, connecting the two transverse furrows. Marginal elytral ridge disappearing posteriorly. 2nd antennal segment short, not enlarged distally.

Ochthebius exaratus Mulsant, 1844
Fig. 52.

Ochthebius exaratus Mulsant, 1844, Hist. Nat. Col. Fr. Palp.: 67.

1.1-1.3 mm. Dark brown to black, dorsal surface shining, without metallic hue. Head and pronotum smooth and impunctate; anterior margin of labrum rounded, not emarginate. Lateral margins of pronotum (anterior to latero-basal excision) rounded (Fig. 52), indistinctly crenulate. Elytra rather convex, and with strongly rounded sides, fairly strongly punctato-striate; each puncture with a very fine curved seta; interstices almost flat, without any trace of reticulation. Appendages yellowish red to yellowish brown, antennal base yellow.

Distribution. So far not recorded from Scandinavia. – Widespread along the coasts of the Mediterranean, the Atlantic coasts of West Europe, north to the English Channel (France and the South of England) and the Netherlands. Perhaps it may be found in south-western Scandinavia.

Biology. A halobiontic species, confined to salt marshes, where it lives in or at the edges of small shallow pools on clayey or sandy ground; it is found in company with e.g. *Ochthebius marinus, O. viridis, Enochrus bicolor, Agabus conspersus* and *Coelambus parallelogrammus*. Only accidentally taken in fresh water. Mainly in spring and autumn.

Ochthebius narentinus Reitter, 1885
Fig. 53.

Ochthebius narentinus Reitter, 1885, Dt. ent. Z. 29: 362.

1.0-1.2 mm. Similar to *exaratus,* but easily recognized by the shape of pronotum (Fig. 53), in which the lateral margins (anterior to latero-basal excision) are strongly angularly emarginate in middle, and broadly bordered by the marginal membrane. Furthermore the elytral punctures are stronger, with longer and more conspicuous setae; the interstices are much narrower, but still almost flat, very indistinctly reticulate.

Distribution. So far not recorded from Scandinavia. – A South-east European species, ranging from northern Italy and the Balkans to Hungary and Czechoslovakia; recently found also in northern West Germany. Perhaps it may also be found in southern Scandinavia.

Biology. Not much is known about the biology of this rare species. Apparently it prefers running water (Ienistea, 1968), but no details are mentioned about its habits. In North Germany is has been taken in flight with a car net (Ziegler, 1984). Adults are taken in July; perhaps the species is most abundant in summer.

Subgenus *Homalochthebius* Kuwert, 1887

Homalochthebius Kuwert, 1887, Dt. ent. Z. 31: 383.
Type-species: *Elophorus pygmaeus* Gyllenhal (sensu Kuwert, 1887 [*nec Elophorus pygmaeus* Fabricius, 1792] = *Elophorus minimus* Fabricius, 1792), by subsequent designation (d'Orchymont, 1942b).

Lateral margins of pronotum distinctly excised in about posterior fourth; the excision only moderately strong (Fig. 49), filled up by the marginal membrane. Pronotum with a conspicuous median longitudinal furrow, but otherwise without distinct depressions on raised middle portion. Marginal elytral ridge disappearing posteriorly.

6. *Ochthebius minimus* (Fabricius, 1792)
Fig. 49; pl. 1: 6.

Elophorus minimus Fabricius, 1792, Ent. Syst. 1 (1): 205.
Hydrophilus impressus Marsham, 1802, Ent. Brit. 1: 408.

1.8-2.1 mm. Black, dorsal surface with a dark bronze hue. Head and pronotum with moderately strong punctuation; anterior margin of labrum not emarginate. Elytra punctato-striate, each puncture with an extremely fine, inconspicuous seta; the punctures becoming finer apically; interstices feebly convex; intercalary striae absent. Appendages yellowish red, antennal base yellow.
♂: Protarsi dilated basally. Outer edge of mandibles with a tuft of short stout setae (as in Fig. 58). Elytra rather shining, only feebly reticulate.
♀: Protarsi and mandibles simple. Elytra rather dull, with very distinct transverse reticulate microsculpture.

Distribution. A common and abundant species, widespread throughout Fennoscandia and Denmark. – Widely distributed throughout most of the Palearctic region, in North and Central Europe the most frequent species of the genus; southwards to the Mediterranean.

Biology. In almost all kinds of fresh water, both stagnant and running, occasionally found also in brackish water. It is usually very abundant, predominantly in shallow water among vegetation; it seems to avoid more oligotrophic waters. An active flyer, often found in drift on the seashore. March-October, mainly in spring and autumn.

Subgenus *Hymenodes* Mulsant, 1844

Hymenodes Mulsant, 1844, Hist. Nat. Col. Fr. Palp.: 68.
Type-species: *Ochthebius pellucidus* Mulsant, 1844 (= *Ochthebius nanus* Stephens, 1829), by present designation.

Lateral margins of pronotum excised in posterior half or more, rather abruptly narrowed, but not as strongly as in *Asiobates;* the excision filled up by the marginal membrane (Fig. 50). Pronotum with a conspicuous median longitudinal furrow, reaching almost from anterior to posterior margin, and with a pattern of depressions very similar to that of *Asiobates,* though these depressions are normally shallower. Marginal elytral ridge very narrow (dorsal view), disappearing posteriorly. Elytra with distinct (regular) series of punctures.

Mulsant included 4 species in his original description of *Hymenodes,* but he did not designate any type-species. One of the species has subsequently been included in another subgenus. Thus, to avoid possible confusion I find it most appropriate to designate *pellucidus* as the type-species.

7. *Ochthebius nilssoni* Hebauer, 1986
Fig. 50.

Ochthebius nilssoni Hebauer, 1986, Ent. scand. 17: 359.

1.6 mm. Black, dorsal surface with distinct metallic hue, on head and pronotum purplish, on elytra more bronze. Head rather dull, only posteriorly (between frontal depressions) more shining. Labrum, clypeus, anterior portion of frons and frontal depressions with very distinct reticulate microsculpture; anterior margin of labrum with a small v-shaped emargination. Pronotum on raised middle portion somewhat shining, with extremely fine and sparse punctuation; towards anterior and posterior margin, and in the depressions distinctly reticulate. Elytra rather narrow, only about 1/4 wider than pronotum, with regular series of fine (apically very fine) punctures; interstices flat, rather shining, only indistinctly reticulate. Entire dorsal surface sparsely covered with rather long, but very thin and little conspicuous hairs. Metasternum reticulate and densely pubescent, without shining field on middle portion. Appendages reddish brown, antennae and maxillary palpi a little paler.

Distribution. Known only from a few specimens, all found at a single locality in North Sweden (Vb.: Degerfors (Skaten, Västra Skärträsket)).

Biology. The few specimens known of this species, are found in shallow water at the edge of a small creek of an oligotrophic lake with sandy (-stony) bottom and growths of e.g. *Lobelia* and *Phragmites*. The lake is extremely deep with very clear and cold water, and with some fluctuations in water level (Anders Nilsson, *in litt.*).

Subgenus *Ochthebius* Leach, 1815, s.str.

Ochthebius Leach, 1815, in Brewst. Edinb. Enc. 9: 95.
   Type species: *Elophorus marinus* Paykull, 1798, by subsequent designation (d'Orchymont, 1942b).

Lateral margins of pronotum only narrowly excised posteriorly, and narrowly bordered by the marginal membrane (Fig. 48). Pronotum without distinct longitudinal furrow, but with two transverse, rather shallow depressions, one before and one behind middle. The lateral longitudinal depression delimiting the raised middle portion of pronotum, shallow or very shallow, sometimes fused with the small, very shallow depression inside posterior angles. Marginal elytral ridge disappearing posteriorly. 2nd antennal segment not enlarged distally.

8. *Ochthebius marinus* (Paykull, 1798)
   Figs. 48, 62, 64, 66; pl. 1 7

Elophorus marinus Paykull, 1798, Fauna Suec. 1: 245.

1.8-2.1 mm. Black, head and pronotum with distinct bronze, or often greenish metallic hue; elytra brown to yellowish brown, often rather pale. Head and pronotum somewhat shining, extremely finely punctuated; clypeus and frons (between frontal depressions and eyes) reticulate; anterior margin of labrum truncate, hardly emarginate. Pronotum reticulate laterally and in the transverse impressions. Elytra rather narrow, rather finely punctato-striate; each puncture with an extremely fine inconspicuous seta; interstices somewhat shining, flat or almost so, with more or less distinct rugulose reticulation. Appendages yellowish red, antennal base normally a little lighter. Legs rather long and slender (Fig. 64).
   ♂: Terminal lobe of aedeagus rather small (Fig. 66).

Distribution. Common and widespread along the coasts of Denmark, southern Norway, Sweden (north to Nb.) and Finland. – Widely distributed along the European coasts, ranging from the White Sea and Fennoscandia to the Mediterranean, from the Atlantic coasts of West Europe to the Caspian Sea; also at saline inland localities throughout most of Europe. Also widely distributed in North America.

Biology. At the edges of almost all kinds of brackish water, but mainly in shallow pools on saltmarshes, where it is very abundant, both in the water and on moist ground round the pools. Though it is clearly halophilic, it apparently tolerates very low salinities, and may occasionally also be taken in fresh water far from the coast. The species is a very active flyer, often found in drift on the seashore, so perhaps records from fresh water are accidental and do not represent stable populations. April-October, most abundant in spring and autumn. Numerous full grown larvae, pupae and newly emerged adults were found at the end of July, in moist soil at the bottom of a dried-up salt marsh pool.

9. *Ochthebius lenensis* Poppius, 1907
   Figs 63, 65.

Ochthebius lenensis Poppius, 1907, Öfvers. finska VetenskSoc. Förh. 49 (2): 10.

1.8-2.0 mm. Very similar to *marinus,* though normally markedly darker. Head and

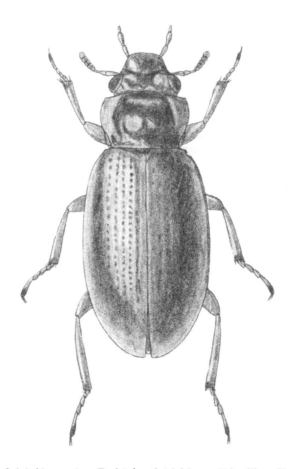

Fig. 62. *Ochthebius marinus* (Payk.), length 1.8-2.1 mm. (After Victor Hansen).

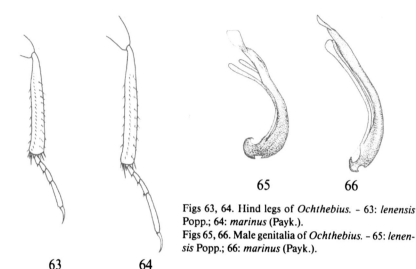

Figs 63, 64. Hind legs of *Ochthebius*. – 63: *lenensis* Popp.; 64: *marinus* (Payk.).

Figs 65, 66. Male genitalia of *Ochthebius*. – 65: *lenensis* Popp.; 66: *marinus* (Payk.).

65   66

63   64

pronotum more dull than in *marinus,* with more distributed reticulation, and darker metallic hue; elytra dark to blackish brown, a little more narrowed posteriorly. Appendages yellowish red, antennal base yellow. Legs, especially the tarsi, shorter than in *marinus* (Fig. 63).

♂: Terminal lobe of aedeagus larger than in *marinus* (Fig. 65).

Distribution. In Scandinavia only known from the northernmost parts of Norway (TRy, Nsi). – Widespread from North Norway and Lapponia rossica eastwards through the northern Palearctis to East Siberia (Yakutsk); also known from a few coastal localities in Scotland.

Biology. The species lives in brackish water, and is taken in shallow pools with rather extensive marshy ground. Beyond this, not much information concerning its biology is available. Adults are found in May-June and September, perhaps most frequently in spring.

10. *Ochthebius viridis* Peyron, 1858
Fig. 57.

Ochtebius viridis Peyron, 1858, Annls Soc. ent. Fr. (3) 6: 404.

1.4-1.6 mm. Black, dorsal surface with a faint bronze hue. Head and pronotum somewhat shining, the punctuation a little stronger than in *marinus.* Clypeus and frons (between frontal depressions and eyes) reticulate; clypeal reticulation sometimes reduced medially; anterior margin of labrum rounded (Fig. 57), without emargination. Pronotum reticulate laterally and in the transverse impressions. Elytral striae a little stronger than in *marinus;* interstices with feeble, but distinct, rugulose reticulation. Appenda-

48

ges yellowish red to yellowish brown, antennal base yellow. Legs shorter than in *marinus*. Metasternum without glabrous, shining field on raised middle portion.

Distribution. In Denmark comparatively rare and sporadic (EJ, WJ, NEJ, LFM, SZ); in Sweden rare, only recorded from Sk. and Öl.; not in Norway and Finland. – Widespread along the coasts of the Mediterranean, also North Africa and Asia Minor; north along the Atlantic coast to the British Isles, the Netherlands, southern Scandinavia and the Baltic coast of North Germany; also at saline inland localities in Europe.

Biology. A halobiontic species, in salt marsh areas along the coasts, where it lives among vegetation at the edges of shallow brackish pools or drainage canals. The species is less euryoecious than *marinus,* but seems to tolerate lower salinities than *auriculatus.* May-October, most abundant in spring and autumn, usually taken in numbers.

Ochthebius pusillus Stephens, 1832
 Fig. 56.

Ochthebius pusillus Stephens, 1832, Ill. Brit. Ent. Mandib. 5: 397.

1.5-1.7 mm. Similar to *viridis,* but recognized from it (as well as the other species of the subgenus) by the very distinct emargination at anterior margin of labrum (Fig. 56). Furthermore the reticulation of head and pronotum is generally weaker and more reduced, clypeus more shining (only reticulate laterally). Raised portion of metasternum more or less shining in middle.

Distribution. So far not recorded from Scandinavia. – South, West and Central Europe, eastwards to the Caucasus, northwards to North Germany (Holstein). Possibly also in southern Scandinavia.

Biology. In stagnant water. The references to the habitat of this species are somewhat conflicting. Probably it mainly inhabits fresh (or almost fresh) water, but there are records from brackish water, though at least some of these may be due to mere repetitions of some old unconfirmed records, probably concerning *viridis.* June, September.

11. *Ochthebius evanescens* J. Sahlberg, 1875
 Pl. 1: 8.

Ochthebius evanescens J. Sahlberg, 1875, Notis. Sällsk. Fauna Flora Fenn. Förh. 14: 208.

1.4-1.5 mm. A very characteristic species, at once recognized by the absence of distinct series of punctures on the elytra. Black, head and pronotum with a distinct purplish bronze or greenish metallic hue, elytra dark and only weakly metallic. Entire dorsal surface reticulate, rather dull, only on raised middle portion of pronotum (outside the impressions) more shining. Head and pronotum impunctate; elytra punctured only in anterior half, the punctures very fine and sparse, somewhat irregularly arranged. Anterior margin of labrum truncate. Appendages yellowish red.

Distribution. An eastern, apparently fairly rare species, recorded from East Fennoscandia (Kr, Lr); not in Denmark, Sweden, Norway and Finland. – Ranges from NW. Russia in the north, and the Balkans in the south, eastwards to Mongolia.

Biology. Not much information about the biology of this species is available. Angus found the species in small slightly brackish pools on the steppe (USSR) (Angus, *in litt.*). According to Ienistea (1968: *O. glabratus*) it is halophilic, and has been found in company with other halophilic *Ochthebius*-spp. (e.g. *viridis*). (The synonymy of *glabratus* and *evanescens* is pointed out by J. Balfour-Browne (Angus, *in litt.*).)

Genus *Hydraena* Kugelann, 1794

Hydraena Kugelann, 1794, *in* Schneid. Mag. 5: 579.
Type species: *Hydraena riparia* Kugelann, 1794, by monotypy.

Body contour interrupted between pronotum and elytra. Head with fairly large, protruding eyes; frons without distinct impressions or furrows; ocelli absent; anterior margin of labrum in middle with a deep and narrow emargination (Fig. 67). Middle portion of pronotum slightly raised, delimited laterally by a more or less distinct longitudinal depression, otherwise without well marked impressions. Shape of pronotum typically more or less markedly hexagonal, normally widest in middle, the margins forming an obtuse (rounded) angle. Pronotum without marginal hyaline membrane. Elytra more or less elongate, normally regularly punctato-striate, the number of striae varying from one species to the other (10-15 striae in the species included here). A rudimentary intercalary stria is sometimes visible at base, between 1st and 2nd stria. Lateral margins of elytra forming a sharp ridge, which in most species disappears apically. Ventral surface of body dull, pronouncedly shagreened, and with dense hydrofuge pubescence. Procoxal cavities closed posteriorly (Fig. 68). Metasternum (Figs 84, 85) medially with two subparallel and mostly rather flat ridges, and between these impres-

67

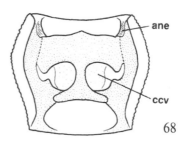

ane

ccv

68

Figs 67, 68. *Hydraena gracilis* Germ. – 67: head; 68: prosternum (legs removed).

sed; the ridges in most species glabrous and shining, well defined. In some species (subgenus *Phothydraena*) the ridges are united anteriorly to a single median ridge, and two similar, but shorter and oblique ridges are present, one on each side of the median ridges. Abdomen in ♂♂ with 6 visible sternites, in ♀♀ with normally 7 visible sternites. Pygidium in ♂♂ often with a more or less pronounced emargination. Maxillary palpi (Figs 71-75) very long and slender, terminal segment longer than penultimate. Antennae 9-segmented with 5-segmented pubescent club. Legs slender and rather long.

A large genus, distributed over all parts of the world, comprising about 400 known species. 10 species occur in Fennoscandia and Denmark or the adjacent areas.

Many species of *Hydraena* show a very poor dispersal ability, particularly the species living in running waters; more of these species are brachypterous or even apterous (Balfour-Browne, 1958) and appear to be very local, being generally very restricted in their areas of distribution. Furthermore, the majority of the species are obviously confined to clear and unpolluted waters, and may serve as indicators for such.

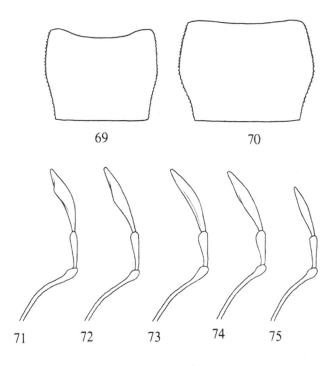

Figs 69, 70. Pronotum of *Hydraena*. - 69: *palustris* Er.; 70: *britteni* Joy.
Figs 71-75. Maxillary palpus of *Hydraena*. - 71: *britteni* Joy, male, 72: *riparia* Kugel., male; 73: *bohemica* Hrbácek, male; 74: *sternalis* Rey, male; 75: *britteni* Joy, female.

1   Elytra apically, along lateral margin, with some very
     large, somewhat transparent, foveate punctures (Fig.
     83)... 12. *testacea* Curtis
  – Elytra without such large punctures.............................. 2
2 (1) Pronotum (Fig. 69) only very weakly narrowed anterior-
     ly, much more narrowed posteriorly, anterior margin
     rather markedly concave .... 13. *palustris* Erichson
  – Pronotum (Fig. 70) more narrowed anteriorly, anterior
     margin only feebly concave...................................... 3
3 (2) Elytra with 15 rather finely and densely punctured, re-
     gular striae, all interstices very narrow, in middle
     generally distinctly narrower than diameter of strial
     punctures... 4
  – Elytra with 10-11 densely punctured, laterally somewhat
     irregular striae, interstices (except laterally) a little wider,
     in middle wider than diameter of strial punctures 10
4 (3) Elytra black, rather short, only about 1/2 longer than
     wide (Fig. 87). Length 1.7-2.0 mm................. 16. *nigrita* Germar
  – Elytra black or brownish, longer, about 2/3 longer than
     wide (Fig. 86). Length 1.9-2.4 mm................................ 5
5 (4) Pronotum rather strongly and closely punctured (except
     on middle portion), distance between punctures towards
     anterior margin generally 1/3 to 1/2 of punctual dia-
     meter. Pronotum with rather extensive reticulation,
     which is distinct both in and between punctures, at least
     on lateral portions and towards anterior and posterior
     margin. ♂: Terminal segment of maxillary palpi asym-
     metrical, inner face with a conspicuous bulge in about
     distal third (Fig. 71) .... 14. *britteni* Joy
  – Pronotum more strongly and very closely punctured (ex-
     cept on middle portion), distance between punctures
     towards anterior margin generally less than 1/4 of punc-

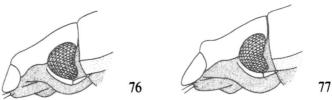

76   77

Figs 76, 77. Head of *Hydraena*-females (lateral view). – 76: *riparia* Kugel.; 77: *sternalis* Rey.

tual diameter. Pronotum at most distinctly reticulate between punctures towards lateral margins. ♂: Terminal segment of maxillary palpi less asymmetrical, the inner face more feebly bulging (or merely finely ridged) (Figs 72-74) 6

6 (5) Inner face of mesotibiae with fine but distinct crenulation in distal half (almost as in fig. 82). Inner face of terminal segment of maxillary palpi finely ridged, and normally distinctly (though sometimes very feebly) bulging in about middle (Figs 72-74) (♂♂) 7

– Inner face of mesotibiae simple, without crenulation. Terminal segment of maxillary palpi simple (as in Fig. 75) (♀♀) 9

7 (6) Metasternal ridges glabrous and shining, very conspicuous and well-defined .... 15. *riparia* Kugelann (♂)

– Metasternal ridges shagreened and dull, thus only vaguely demarcated from rest of metasternum 8

8 (7) Terminal segment of maxillary palpi a little asymmetrical, slightly curved inwards (Fig. 73). Aedeagus as in Fig. 90 .... *bohemica* Hrbácek (♂)

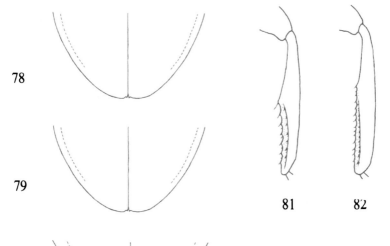

78

79

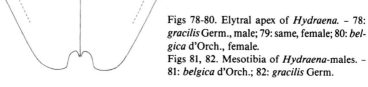

81   82

80

Figs 78-80. Elytral apex of *Hydraena*. – 78: *gracilis* Germ., male; 79: same, female; 80: *belgica* d'Orch., female.
Figs 81, 82. Mesotibia of *Hydraena*-males. – 81: *belgica* d'Orch.; 82: *gracilis* Germ.

53

–　　　　Terminal segment of maxillary palpi more symmetrical,
　　　　not curved inwards (Fig. 74). Aedeagus as in Fig. 91 . . . . *sternalis* Rey (♂)
9 (6)　Clypeus with a rather strong and conspicuous antero-
　　　　median bulge (Fig. 77) . . . . *sternalis* Rey (♀)
–　　　　Clypeus less strongly bulging (Fig. 76) . . . .
　　　　. . . . . 15. *riparia* Kugelann (♀) and *bohemica* Hrbácek (♀)
10 (3)　Smaller, 1.5-1.6 mm. Elytra and pronotum, except for
　　　　a dark transverse band on the latter, yellowish brown . .
　　　　. . . . . 17. *pulchella* Germar
–　　　　Larger, 2.1-2.4 mm. Elytra and pronotum uniformly
　　　　piceous to black 11
11 (10)　Inner face of mesotibiae crenulate in distal half (Figs 81,
　　　　82); inner face of metatibiae with a fringe of long stiff
　　　　hairs in about distal half. Elytral apex smoothly round-
　　　　ed, not excised (Fig. 78) (♂♂) 12
–　　　　Meso- and metatibia simple. Elytral apex with a distinct
　　　　semicircular excision (Figs 79, 80) (♀♀) 13
12 (11)　Inner face of mesotibiae straight in distal half, the cre-
　　　　nulation very fine (Fig. 82). Aedeagus as in Fig. 93. . . . .
　　　　. . . . . 18. *gracilis* Germar (♂)
–　　　　Inner face of mesotibiae feebly convex in distal half, the

Fig. 83. Elytral apex of *Hydraena testacea*
Curt.
Figs 84, 85. Metasternum of *Hydraena*. – 84:
testacea Curt.: 85: *palustris* Er.

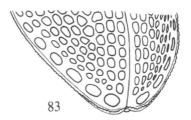

83

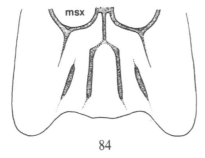

84

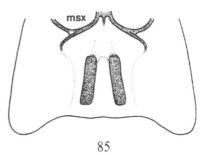

85

crenulation a little stronger (Fig. 81). Aedeagus as in
Fig. 94 .... *belgica* d'Orchymont (♂)
13 (11) Apical excision of elytra rather small, each elytral apex
rather sharply angled (Fig. 79) .... 18. *gracilis* Germar (♀)
  –   Apical excision of elytra much larger, each elytral apex
rounded (Fig. 80) .... *belgica* d'Orchymont (♀)

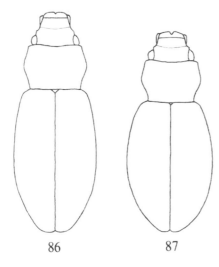

86   87

Figs 86, 87. Outline of *Hydraena.* – 86: *britteni* Joy; 87: *nigrita* Germ.

Subgenus *Phothydraena* Kuwert, 1888

Phothydraena Kuwert, 1888, Dt. ent. Z. 32: 114.
    Type species: *Hydraena testacea* Curtis, 1830, by monotypy.

Elytra apically with some very large transparent, foveate punctures along the lateral margin (Fig. 83). Pseudepipleura broadly visible to apex. Metasternum (Fig. 84) with 4 glabrous ridges; the two medial ridges rather long, slightly converging anteriorly, where they unite to form a single median ridge; the two lateral ridges shorter and oblique, converging anteriorly. Male genitalia with parameres.

12. *Hydraena testacea* Curtis, 1830
    Figs 83, 84; pl. 1: 1.

Hydraena testacea Curtis, 1830, Brit. Ent. 7: no. 307.

1.8-2.0 mm. Black; pronotum and elytra yellowish brown, the former with a very broad darker transverse band in middle. Dorsal surface rather dull. Frons moderately strongly and very densely punctured; clypeus without distinct punctuation, markedly reticulate. Pronotum almost as long as wide, widest just behind middle, very weakly narrowed anteriorly, more strongly narrowed posteriorly, lateral margins slightly sinuate posteriorly; anterior margin rather markedly concave; pronotal punctuation strong and very close. Elytra with 12 strongly and extremely closely punctuated striae, interstices very narrow and almost ridged, only about 1/4 as wide as diameter of strial punctures, the punctures of 12th stria greatly enlarged apically (Fig. 83), foveate and semitransparent. Appendages yellowish red.

Distribution. Very rare in Scandinavia; in Denmark only a few localities in SJ, WJ and F; from Sweden only a few specimens from Sm. and Vg.; not in Norway and Finland. – Widespread in West Europe, including the British Isles; from France and Spain eastwards to North Italy, Austria, West Germany and Scandinavia.

Biology. The species lives in stagnant fresh water; there are a few – probably accidental – records from running water. Usually found singly or only few in number, but occasionally abundant; in Denmark found numerously in a small eutrophic, grassy and muddy, somewhat shaded pond, rich in *Lemna.* May, October.

Subgenus *Hydraena* Kugelann, 1794, s. str.

Hydraena Kugelann, 1794, *in* Schneid. Mag. 5: 579.
   Type species: *Hydraena riparia* Kugelann, 1794, by monotypy.

Elytral striae uniform, lateral punctures not enlarged apically. Pseudepipleura often broad, but strongly narrowed posteriorly, not reaching elytral apex. Metasternum (Fig. 85) medially with two rather long, anteriorly slightly converging and (except for ♂♂ of a few species) glabrous and shining ridges (corresponding to inner metasternal ridges in *Phothydraena*). Elytra – in the here treated species – with 15, or (*pulchella*) 10 striae. Male genitalia with parameres.

13. *Hydraena palustris* Erichson, 1837
    Figs 69, 85.

Hydraena palustris Erichson, 1837, Käf. Mark Brand. 1: 200.

1.6-1.8 mm. Black; pronotum and elytra yellowish brown, the former in middle with a very broad dark transverse band across whole width. Frons closely and not very strongly punctured; clypeus reticulate, without distinct punctuation. Pronotum (Fig. 69) a little wider than long, widest just behind middle, very weakly narrowed anteriorly, more strongly narrowed posteriorly, lateral margins slightly sinuate or almost straight posteriorly; anterior margin rather markedly concave. Pronotal punctuation very close and rather strong. Elytra with rather finely and densely punctuated striae, interstices very narrow, generally slightly narrower than strial punctures, rather flat. Metasternal ridges rather wide, glabrous in both sexes. Appendages yellowish red.

Distribution. A comparatively uncommon species, towards the north becoming rare or very rare; found in most provinces of Denmark and southern Sweden, north to Vrm. and Upl., southernmost parts of Norway (AK, HEs, Bø, VAi), and southern Finland (Ab, Ta, Sb, Kb). – Ranges from Scandinavia, England, North and East France to Spain and Italy, eastwards to the northern Balkans, Czechoslovakia and Poland.

Biology. Bound to the edges of stagnant fresh water, often in woodland. It appears to be a typical inhabitant of clear shallow water, living among swampy moss, but not infrequently sieved from leaf litter and the like on moist or wet, mirey ground. April-June, August-October, mainly found in spring.

14. *Hydraena britteni* Joy, 1907
  Figs 70, 71, 75, 86, 88; pl. 1: 3.

Hydraena britteni Joy, 1907, Entomologist's mon. Mag. 43: 79.

1.9-2.2 mm. Brown to black; dorsal surface rather dull. Frons rather finely and densely punctured; clypeus very finely punctured and (at least laterally) reticulate. Pronotum (Fig. 70) about 1/3 wider than long, widest in middle, distinctly narrowed both anteriorly and posteriorly, lateral margins straight or slightly sinuate posteriorly; anterior margin only feebly concave. Pronotum rather strongly and closely punctured, distance between punctures towards anterior margin generally 1/3 to 1/2 of puncture diameter; also with fairly extensive reticulation, which is distinct at least on lateral portions, and towards anterior and posterior margin, both in and between the punctures. Middle portion of pronotum however rather shining, finer and more sparsely punctured, without or only with obsolete reticulation. Elytral striae rather finely and densely punctured (strongest anteriorly), interstices very narrow, distinctly narrower than diameter of strial punctures (except posteriorly), rather flat, indistinctly reticulate. Elytra rather narrow, almost 2/3 longer than wide. Metasternal ridges glabrous in both sexes. Appendages yellowish red, antennal base yellow.

  ♂: Terminal segment of maxillary palpi asymmetrical (Fig. 71), curved inwards, inner face with a conspicuous bulge in about distal third. Inner face of mesotibiae with fine crenulation in distal half. Aedeagus as in Fig. 88.

  ♀: Terminal segment of maxillary palpi simple (Fig. 75). Mesotibiae simple.

Distribution. A fairly common species, widely distributed throughout Denmark and Fennoscandia. – North and Central Europe, southward to southern Germany (Bavaria); ranges from the British Isles and the North of France to northern Russia.

Biology. The species lives in or at the edges of fresh, predominantly stagnant waters; mainly in more or less eutrophic, shallow, somewhat shaded pools among moss, leaf litter, or the like, often in woodland. Sometimes at the same sites as *palustris,* but more euryoecious, and occasionally found also in running water, at the grassy edges or under stones in the water. March-October, predominantly in spring, but also abundant in late summer and autumn.

15. *Hydraena riparia* Kugelann, 1794
Figs 72, 76, 89.

Hydraena riparia Kugelann, 1794, *in* Schneid. Mag. 5: 579.

2.1-2.3 mm. Very similar to *britteni,* but pronotum, especially towards anterior and posterior margin with stronger and extremely close, almost rugose punctuation, distance between punctures towards anterior margin generally less than 1/4 of puncture diameter; pronotal reticulation reduced, at most visible between the punctures towards lateral margins. Lateral margins of pronotum on the average a little stronger sinuate posteriorly, and elytral punctures usually larger than in *britteni.* Metasternal ridges glabrous in both sexes.

♂: Terminal segment of maxillary palpi slightly asymmetrical (Fig. 72), curved slightly inwards, inner face finely ridged and feebly bulging in (or just distal to) middle. Inner face of mesotibiae with fine crenulation in distal half. Aedeagus as in Fig. 89.

♀: Terminal segment of maxillary palpi simple. Mesotibiae simple. Clypeus only slightly bulging antero-medially (Fig. 76).

Distribution. Widespread in Scandinavia, but generally less frequent than *britteni;* northward to NTi in Norway, Lu. Lpm. and Nb. in Sweden, and Om in Finland. – Ranges from Scandinavia to Central Spain, Central Italy and the Bosporus, eastward to the Caucasus.

Biology. Predominantly in clear, unpolluted, running water, but also taken among detritus at the edges of lakes; it is not found in small stagnant pools (which are typical of *britteni*). It is usually found under stones or among shingle or gravel on the bottom of rivers or streams, as well as on submerged branches or twigs. April-October, mainly in spring.

Hydraena bohemica Hrbácek, 1951
Figs 73, 90.

Hydraena bohemica Hrbácek, 1951. Čas. čsl. Spol. ent. 48: 223.

2.1-2.3 mm. Very much resembling *riparia,* differing only in the sexual characters of male.

♂: Metasternal ridges shagreened and dull as rest of metasternum, rather feeble and thus only vaguely demarcated. Terminal segment of maxillary palpi curved slightly inwards (Fig. 73), the inner face finely ridged, but hardly bulging in middle. Mesotibia as in *riparia.* Aedeagus as in Fig. 90.

♀: As *riparia-* ♀, i.e. metasternal ridges glabrous and clypeus only slightly bulging antero-medially.

Distribution. So far not recorded from Scandinavia. – Europe; the distribution is still not known in details (see also under *sternalis*). Described from Czechoslovakia;

from here ranging southwards to the Balkans (possibly widespread in the Mediterranean and in East Europe); westwards at least to northern Switzerland and Austria, widespread in Central Europe, where it is found north to Oldenburg and Holstein (near Kiel). Probably also in southern Scandinavia.

Biology. In clear, unpolluted, running water, especially in mountainous areas; in the North German lowland it has been found at the edge of a lake under immersed pieces of wood and among detritus (together with the dytiscids *Stictotarsus duodecim-pustulatus* and *Platambus maculatus*), and in deciduous woodland, in a pool with springy edges. Mainly found in spring and early summer.

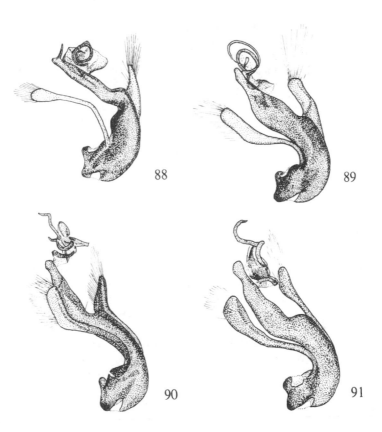

Figs 88-91. Male 88-91. Male genitalia of *Hydraena*. – 88: *britteni* Joy; 89: *riparia* Kugel.; 90: *bohemica* Hrbácek; 91: *sternalis* Rey.

Hydraena sternalis Rey, 1893
Figs 74, 77, 91.

Hydraena sternalis Rey, 1893, Bull. Soc. ent. Fr. 62: 9.

2.2-2.4 mm. Very much resembling *riparia* and *bohemica,* differing only in the sexual characters.

♂: Metasternal ridges (as in *bohemica*) dull and feeble, rather inconspicuous. Terminal segment of maxillary palpi almost symmetrical (Fig. 74), the inner face finely ridged, and very weakly bulging in middle. Mesotibiae as in *riparia.* Aedeagus as in Fig. 91.

♀: Metasternal ridges glabrous and well-defined. Terminal segment of maxillary palpi simple. Mesotibiae simple. Clypeus strongly bulging antero-medially (Fig. 77), stronger than in both *riparia* and *bohemica.*

Distribution. So far not recorded from Scandinavia. – Europe; the species has been confused with *bohemica,* and its distribution is still incompletely known. Apparently it is more north-westerly than *bohemica. H. sternalis* is recorded from Central and North France, Belgium and West Germany (e.g. Holstein); records of *sternalis* from eastern and south-eastern Europe probably mostly concern *bohemica. H. sternalis* may also be found in South Scandinavia.

Biology. Probably similar to *riparia* and *bohemica,* i.e. preferring clear and unpolluted running waters. Lohse (*in litt.*) has taken it in moss slightly wetted by water at a small river (where it occurred together with *Elmis maugetii* Latr.).

16. *Hydraena nigrita* Germar, 1824
Fig. 87.

Hydraena nigrita Germar, 1824, Ins. spec. nov.: 93.

1.7-2.0 mm. A black, rather short species (Fig. 87). Dorsal surface rather dull. Head densely and finely (on clypeus very finely) punctured; clypeus reticulate, rather dull, except medially, where reticulation is more indistinct. Pronotum widest at about middle, distinctly narrowed both anteriorly and posteriorly, lateral margins slightly sinuate posteriorly; anterior margin only feebly concave; pronotal punctuation rather strong and close, reticulation in ♀ often rather extensive (almost as in *britteni*), in ♂ usually weaker. Elytra shorter than in *britteni* and *riparia,* only 1/2 longer than wide, with rather finely and densely punctured striae, punctures becoming distinctly smaller in posterior half; interstices narrow, anteriorly a little narrower than diameter of strial punctures, posteriorly almost as wide as the punctures, flat. Metasternal ridges glabrous in both sexes. ♂ without distinct characteristics on palpi and tibiae. Appendages yellowish red.

Distribution. Comparatively rare in Denmark (SJ, EJ, SZ, NEZ, B); in Norway only recorded from a few provinces (AK, TEy, VAy); recently found also in Sweden at a few localities in Sk. (Dalby and between Harlösa and Öveds Kloster; L. Huggert *in litt.*); not in Finland. – West and Central Europe, ranging from the British Isles, North and Central France to Hungary, Czechoslocakia, Poland and the Baltic States of the USSR; from southern Scandinavia to North Italy and the northern Balkans.

Biology. An exclusive inhabitant of clear, unpolluted, running water, mainly under stones or among gravel or shingle on the bottom, or sitting on submerged branches and the like; often in small woodland streams. April-October, mainly in spring and early summer, and in autumn.

17. *Hydraena pulchella* Germar, 1824
Pl. 1: 4.

Hydraena pulchella Germar, 1824, Ins. spec. nov.: 94.

1.5-1.6 mm. Black; pronotum and elytra yellowish brown, the former with a very broad dark transverse band across the whole width. Head somewhat shining, posteriorly rather finely and not very densely punctuated; clypeus reticulate at least laterally. Pronotum widest in middle, rather strongly narrowed both anteriorly and posteriorly, in almost straight or (posteriorly) sometimes slightly sinuate lines; anterior margin only feebly concave; pronotal punctuation dense and moderately strong, on middle a little finer. Elytra (contrary to the other species of the subgenus) with only 10, laterally somewhat irregular series of punctures; the punctuation fairly strong anteriorly, much finer posteriorly; interstices almost flat, in anterior third (and laterally) a little narrower than width of strial punctures, otherwise distinctly wider than the punctures. Metasternal ridges glabrous in both sexes. Appendages yellowish red.

Distribution. A very rare species; in Denmark known only from a few localities in Jutland and Funen (EJ, WJ, NEJ, F); in Sweden only few sporadic findings from Sk., Hall. and Upl.; in Finland mainly in the southern provinces (Ab, N, Ta, Sa, Kb, ObN); not in Norway. – Widespread in Central Europe, ranging from Fennoscandia to North Italy, North Yugoslavia and Bulgaria, from France and the British Isles to Czechoslovakia, Poland and NW. Russia.

Biology. The species lives in clear unpolluted running water, mainly at grassy edges of rivers and streams, only occasionally under stones or in moss in the water. Though it may be taken in numbers, the populations are very local and apparently very static, probably due to a poor ability of dispersal. June-July.

Subgenus *Haenydra* Rey, 1886

Haenydra Rey, 1886, Annls Soc. linn. Lyon 32 (1885): 95.
Type species: *Hydraena lapidicola* Kiesenwetter, 1849, by subsequent designation (d'Orchymont, 1936a).

Elytra with 11, laterally somewhat irregular and close-set, punctuated striae; lateral punctures not enlarged apically. Metasternum (as in *Hydraena* s.str.) with only two ridges, these always glabrous. Males are characteristic by the absence of parameres in the genitalia.

18. *Hydraena gracilis* Germar, 1824
Figs 67, 68, 78, 79, 82, 92, 93; pl. 1: 2.

Hydraena gracilis Germar, 1824, Ins. spec. nov.: 94.

2.1-2.3 mm. A very elongate species (Fig. 92). Piceous to black, dorsal surface somewhat shining. Head posteriorly moderately strongly and rather densely punctuated; clypeus very finely punctured and feebly reticulate. Pronotum a little wider than long, widest in middle, rather strongly narrowed both anteriorly and posteriorly, in straight or (posteriorly) often slightly sinuate lines; anterior margin only feebly concave. Pronotum with dense and moderately strong punctuation, and indistinct reticulation;

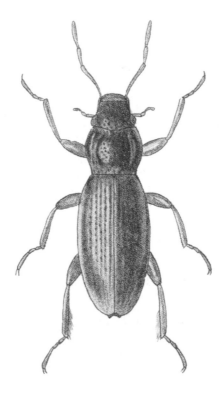

Fig. 92. *Hydraena gracilis* Germ., length 2.1-2.3 mm. (After Victor Hansen).

middle portion shining, finer and sparser, usually rather unevenly punctured. Elytra long and subparallel, about 3/4 longer than wide, with rather fine, densely punctuated striae, that laterally become somewhat irregular and more close-set; interstices flat and (except laterally) wider than in the preceeding species, about 1½ times as wide as width of strial punctures. Lateral flattened margins of elytra strongly narrowed apically in both sexes. Antennae and maxillary palpi yellowish red, legs slightly darker.

♂: Inner face of mesotibiae straight in about distal half, with very fine crenulation (Fig. 82); inner face of metatibiae with a fringe of fairly long stiff hairs in about distal half. Elytral apex smoothly rounded, not excised (Fig. 78). Aedeagus as in Fig. 93.

♀: Tibiae simple. Elytral apex with a small excision, the two elytra together showing a small semicircular gap (Fig. 79).

Distribution. Widespread throughout most of Denmark and Fennoscandia, generally rather frequent, yet from Finland more sparse records. – North and Central Europe, most abundant in mountainous regions, ranging from the British Isles and Fennoscandia to Spain, Italy and northern Yugoslavia, from France to Hungary, Poland and NW. Russia.

Biology. In running, comparatively clear and unpolluted water, under stones or in moss on them, and on submerged branches, twigs, etc., or among detritus, often together with *riparia* and *nigrita*. Apparently most abundant in summer, but also found in spring.

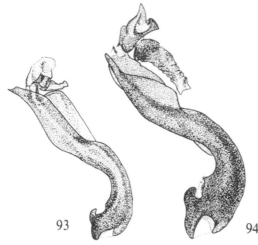

Figs 93, 94. Male genitalia of *Hydraena*. – 93: *gracilis* Germ.: 94: *belgica* d'Orch.

93

94

Hydraena belgica d'Orchymont, 1929
   Figs 80, 81, 94.

Hydraena belgica d'Orchymont, 1929b, Bull. Annls Soc. ent. Belg. 69: 373.

2.1-2.4 mm. Very much resembling *gracilis,* differing only in the sexual characters.

   ♂: Inner face of mesotibiae feebly concave and finely crenulate, the crenulation a little stronger than in *gracilis.* Elytral apex smoothly rounded, not excised (as Fig. 78). Lateral flattened margins of elytra strongly narrowed apically. Aedeagus as in Fig. 94.

   ♀: Tibiae simple. Elytral apex pronouncedly excised, the two elytra together showing a rather large semicircular gap (Fig. 80), elytral apices separately rounded (not angular, as in *gracilis*).

   Distribution. So far not recorded from Scandinavia. – A mainly mountainous species, ranging from eastern France and Belgium to East Europe, where it is widespread; from North Italy and Yugoslavia to North Germany (Holstein). Perhaps it may also be found in South Scandinavia.

   Biology. The species inhabits clear streams, predominantly in mountainous areas. Only few details about its biology are available, but probably it is similar to that of *gracilis.* Perhaps mainly in spring.

Genus *Limnebius* Leach, 1815

Limnebius Leach, 1815, *in* Brewst. Edinb. Enc. 9: 96.
   Type species: *Hydrophilus nitidus* Marsham, 1802, by monotypy.

Body contour evenly curved, not interrupted between pronotum and elytra. Head rather short, eyes only feebly convex; ocelli absent; anterior margin of labrum truncate sometimes distinctly emarginate in middle. Pronotum wider than long, widest at base,

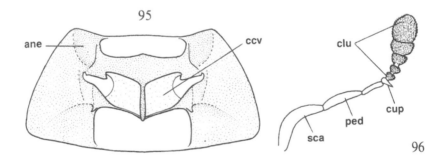

Fig. 95. Prosternum of *Limnebius crinifer* Rey (legs removed).
Fig. 96. Right antenna of *Limnebius truncatulus* Thoms.

strongly narrowed anteriorly, regularly convex, without any trace of impressions. Elytra widest anterior to middle, more or less strongly narrowed posteriorly, truncate at apex. Dorsal surface with fine or very fine reticulate microsculpture, which may be reduced, especially on head and pronotum; also with fine or very fine (in some species hardly visible) punctuation. Ventral surface of body rather dull, pronouncedly shagreened and with dense hydrofuge pubescence. Procoxal cavities closed posteriorly by a narrow chitinous bridge (Fig. 95). Abdomen with 7 visible sternites, the last two

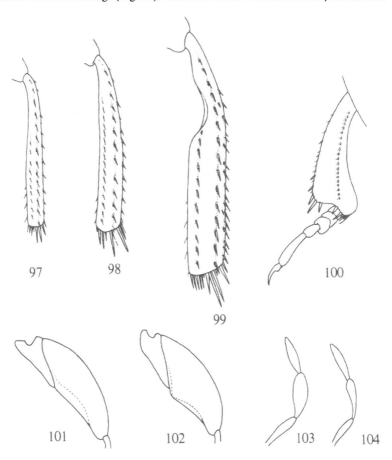

Figs 97-99. Metatibiae of *Limnebius*. – 97: *crinifer* Rey, female; 98: *truncatellus* (Thbg.), female; 99: same, male.

Fig. 100. Protibia and -tarsus of *Limnebius truncatellus* (Thbg.), male.

Figs 101, 102. Metafemora of *Limnebius*-males. – 101: *papposus* Muls.; 102: *truncatulus* Thoms.

Figs 103, 104. Maxillary palpus of *Limnebius*-males. – 103: *papposus* Muls.; 104: *crinifer* Rey.

(at least partially) more or less shining, and at most sparsely pubescent. 6th sternite in
♂ ♂ very large (Figs 106-110), to some extent concealing 7th sternite, and in some species with a large patch of long stiff setae, or with depressions, bulges etc.; in ♀ ♀ of normal size (Fig. 111), not considerably differing in size from the preceeding sternites. Pygidium with two apical, more or less widely separated tufts of long fine hairs. Maxillary palpi fairly long and slender, terminal segment as long as penultimate. Antennae (Fig. 96) 9-segmented with 5-segmented pubescent club. Legs moderately long; tarsi, particularly pro- and mesotarsi, in ♂ dilated basally. In addition to the tarsal and abdominal characters, the ♂ ♂ of *Limnebius* often exhibit other characteristics, such as strongly enlarged tibiae, dilated palpi etc.; in some species are ♂ ♂ considerably larger than ♀ ♀.

The genus comprises about 100 known species, represented in the Palearctic, Nearctic and Oriental regions, mainly in the temperate zones. The species are generally very uniform, showing much less diversity than in the two preceeding genera. 7 species occur in North Europe.

Key to species of *Limnebius*

1   Very small species, 1.0-1-3 mm. Head and pronotum at
      most indistinctly punctured................................... 2
–   Markedly larger species, 1.7-2.8 mm, or (*nitidus:* 1.4-1.7
      mm) pronotum very distinctly punctured 3
2 (1) Entire dorsal surface with fine, but quite distinct reticulation .... 24. *aluta* Bedel
–   Head and pronotum shining, without reticulation .... *atomus* (Duftschmid)
3 (1) Smaller, 1.4-1.7 mm. Pronotum finely, but very distinctly
      punctured, in middle shining, without reticulation ... 23. *nitidus* (Marsham)
–   Larger, 1.7-2.8 mm. Pronotum also in middle distinctly
      reticulate (except in *truncatellus:* >1.9 mm) 4
4 (3) Pronotum finely, but very distinctly punctured, in middle
      shining, at most very indistinctly reticulate. ♂: Pro- and
      mesotibiae strongly curved (Fig. 100); metatibiae strongly
      and abruptly enlarged in distal 2/3 (Fig. 99). ♀: Tibiae
      only weakly curved, not enlarged, yet fairly stout (Fig. 98)...............
      .... 19. *truncatellus* (Thunberg)

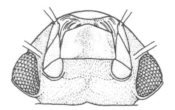

Fig. 105. Ventral surface of head of *Limnebius papposus* Muls.

66

– Pronotum very finely, or even indistinctly punctured, also in middle with distinct, though very fine reticulate microsculpture. Meso- and metatibiae in both sexes rather slender (as in Fig. 97) 5

5 (4) Pronotum hardly punctured in middle, but very distinctly reticulate. ♂: Posterior margin of metafemora angular

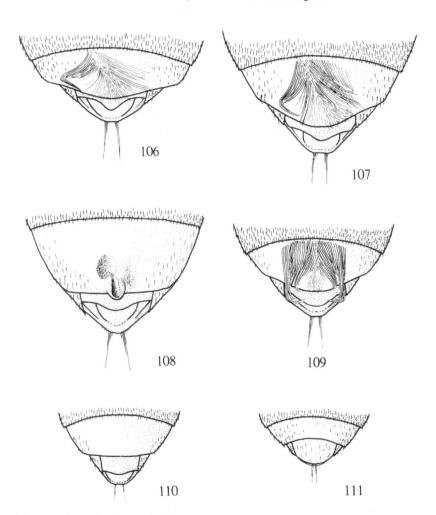

Figs 106-111. Last abdominal sternites of *Limnebius*. – 106: *papposus* Muls., male; 107: *crinifer* Rey, male; 108: *truncatellus* (Thbg.), male; 109: *truncatulus* Thoms., male; 110: *nitidus* (Marsh.), male; 111: same, female.

(Fig. 102); 6th visible abdominal sternite with a large symmetrical patch of long stiff setae (Fig. 109) . . . . 22. *truncatulus* Thomson
- Pronotum in middle with very fine punctuation besides very fine reticulation. ♂: Posterior margin of metafemora not angular (Fig. 101); patch of long stiff setae on 6th sternite strongly asymmetrical (Figs 106, 107) 6
6 (5) Pronotum and elytra yellowish brown, the former with a large well defined black spot in middle. Mentum in its whole length with a deep median depression (Fig. 105). ♂: 3rd segment of maxillary palpi strongly dilated (Fig. 103) . . . . 20. *papposus* Mulsant
- Pronotum and elytra black, except in immature individuals; pale immature specimens without well defined black middle spot on pronotum. Mentum flat or only feebly concave. ♂: Maxillary palpi without dilated 3rd segment (Fig. 104) . . . . 21. *crinifer* Rey

Subgenus *Limnebius* Leach, 1815, s.str.

Limnebius Leach, 1815, *in* Brewst. Edinb. Enc. 9: 96.
   Type species: *Hydrophilus nitidus* Marsham, 1802, by monotypy.
(Other names given by Kuwert 1890, e.g. *Tricholimnebius* and *Embololimnebius,* are superfluous).

Elytra without, or at most with an extremely fine or indistinct sutural stria apically. Pygidial hairtufts (in the species treated here) separate, in ♀ narrowly, in ♂ more widely separated (Figs 110, 111). Male genitalia with parameres, often asymmetrical and quite complex (Figs 113, 114).

19. *Limnebius truncatellus* (Thunberg, 1794)
   Figs 98-100, 108; pl. 1: 10.

Hydrophilus truncatellus Thunberg, 1794, Diss. Ins. Suec. 6: 73.

1.9-2.8 mm. Black; pronotal and elytral margins reddish. Head and pronotum rather shining, with fine, moderately dense punctuation, and very fine reticulation, which is normally absent (or at most very indistinct) on middle portion of pronotum, so the punctuation here appears very distinct. Anterior margin of labrum truncate, emarginate in middle. Elytra somewhat shining, finely and evenly reticulate, and finely, moderately densely punctured; punctuation quite distinct in anterior half, becoming weaker posteriorly. Appendages reddish brown, femora darker, antennal base paler.
   ♂: 2.5-2.8 mm. Pro- and mesotibiae strongly curved inwards, enlarging distally (Fig. 100). Metatibiae (Fig. 99) strongly and abruptly enlarged in about distal 2/3. 6th visible abdominal sternite with a postero-medial (somewhat asymmetrical) tubercular

bulge and some asymmetrical depressions, without long setae (Fig. 108). Maxillary palpi not dilated.

♀: 1.9-2.2 mm. Pro- and mesotibiae much less curved and enlarged distally; metatibiae (Fig. 98) almost straight, not strongly and abruptly enlarged, yet more stout than in ♀♀ (and ♂♂) of our other species.

Distribution. A very widespread and common species, found in the whole of Denmark, Sweden and Norway; in Finland north to Om. – North and Central Europe, widely distributed, south to Central Spain and North Italy, from the British Isles eastwards to the USSR (? Siberia).

Biology. In almost all kinds of running water, mainly found in shallow water near the edges among grasses, mosses etc.; also, but less frequently, in stagnant water. Most abundant in spring and again in late summer and autumn; newly emerged specimens are found at the end of July. Females are normally far more numerous than males.

20. *Limnebius papposus* Mulsant, 1844
Figs 101, 103, 105, 106; pl. 1: 11.

Limnebius papposus Mulsant, 1844, Hist. Nat. Col. Fr. Palp.: 92.

1.8-2.4 mm. Black; pronotum and elytra yellowish brown, the former with a large, well defined median black spot (the presence of this spot usually separates this species from immature individuals of other similar species). Dorsal surface rather shining, with extremely fine, rather dense punctuation, and very fine reticulation, which is distinct also on middle portion of pronotum; the punctuation here extremely fine, but visible; on the elytra rather indistinct. Anterior margin of labrum only feebly emarginate. Mentum (Fig. 105) in its whole length with a deep median depression (in our other species almost flat or much more feebly depressed, and not for entire length). Appendages yellowish red, femora darker basally.

♂: 2.0-2.4 mm. 3rd segment of maxillary palpi strongly dilated, much thicker than 2nd and 4th (Fig. 103). Legs rather slender, metatibia fairly thin, widest in middle. Posterior margin of metafemora not angular (Fig. 101). 6th visible abdominal sternite with a large, strongly asymmetrical patch of long stiff setae (Fig. 106).

♀: 1.8-2.1 mm. Maxillary palpi simple; metatibiae slightly thicker (as Fig. 97).

Distribution. Very rare, in Scandinavia only known from Denmark and South Sweden; not in Norway and Finland. Formerly, it was not uncommon in southern and eastern Denmark, but since the 1920's there are no records. Only a few reliable specimens are known from Sweden (Sk., Vg.), none of which are of recent date. – Widely distributed throughout Central and South Europe; from France (except the Mediterranean area) and SE. England through North and Central Italy to Bulgaria, Asia Minor and the Caucasus; in Central Europe most abundant towards the east.

Biology. The species lives in fresh stagnant water, probably preferring open and fairly warm, clear eutrophic ponds, where it lives among vegetation in shallow water near the edges. Also taken in sea drift (T. Palm, *in litt.*). Mainly found in spring or early summer.

Note. This species has been the subject of some confusion. Examination of a large material of *Limnebius*-spp. showed that many of the specimens standing as *L. papposus* (and all specimens of comparatively recent date) must be classed with (immature individuals of) *crinifer, truncatulus* or even *truncatellus*. This situation has clearly been overlooked, when Hansen as late as 1964 refers to *papposus* as common in Denmark. No doubt owing to the many misidentifications, and to the fact that Thomson, when dealing with the Swedish species, obviously included *crinifer* under *papposus*, the true *papposus* has (also in Sweden) been considered more widespread than it really is. Thus, a number of Swedish provinces have been given, but only a few specimens from Sk. and Vg. could be confirmed. Therefore, records from all other provinces (as mentioned by Lindroth, 1960) are omitted here.

21. *Limnebius crinifer* Rey, 1885
Figs 95, 97, 104, 107.

Limnebius crinifer Rey, 1885, Annls Soc. linn. Lyon 31 (1884): 15.

1.7-2.3 mm. Black, pronotal and elytral margins reddish. Dorsal microsculpture as in *papposus*. Thus, apart from the colour (and the ♂-characters) extremely similar to this, and only distinguished by the much more weakly depressed (or even flat) mentum. Appendages yellowish to reddish brown, femora darker, antennal base yellow.

♂: 2.1-2.3 mm. Legs rather slender, metatibiae fairly thin (as in *papposus*). Posterior margin of metafemora not angular. 6th abdominal sternite almost as in *papposus* (Fig. 107). Maxillary palpi not dilated (though slightly thicker than in ♀) (Fig. 104).

♀: 1.7-2.0 mm. Metatibiae slightly thicker than in ♂ (Fig. 97).

Distribution. Fairly common in Denmark; widespread in South Sweden north to Vrm. and Gstr.; in Finland sporadic and rare, only mentioned from N, Ta, Sb and Kb; not recorded from Norway. – Ranges from Fennoscandia to Switzerland and Austria, from England, the Netherlands and Belgium to Hungary, Czechoslovakia and NW. Russia; in Central Europe most frequent towards the north and the east.

Biology. Mainly inhabiting running water, where it lives in shallow water among vegetation or in wet mud at the edges, particularly found in open, rather soft-bottomed streams and in drainage canals; also, but much less frequently, in stagnant water. Mainly in spring and autumn.

22. *Limnebius truncatulus* Thomson, 1853
Figs 96, 102, 109, 112, 113.

Limnebius truncatulus Thomson, 1853, Öfvers. K. VetenskAkad. Förh. 10: 48.

1.7-2.3 mm. Very similar to *crinifer*, but recognized by the more distinct reticulation on

70

the dorsal surface (strongest in ♀). The punctuation of head, pronotum and elytra strongly reduced, and at most very indistinct; on middle portion of pronotum normally not detectable. Colour as in *crinifer*. Mentum flat or only weakly depressed.

♂: 2.1-2.3 mm. Metatibia fairly thin, widest in middle (almost as in *papposus* and *crinifer*). Posterior margin of metafemora distinctly angular (Fig. 102). 6th visible abdominal sternite with a large symmetrical patch of long stiff setae (Fig. 109). Maxillary palpi not dilated. Aedeagus, Fig. 113.

♀: 1.7-2.0 mm. Metatibiae slightly thicker than in ♂.

Distribution. Widespread in South and Central Scandinavia; fairly common in

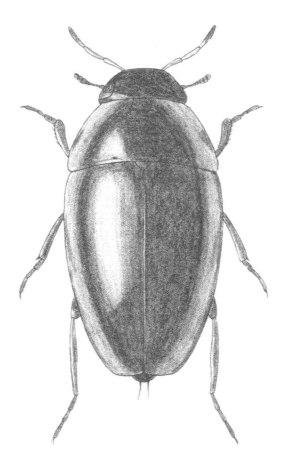

Figs 112. *Limnebius truncatulus* Thoms., length 1.7-2.3 mm. (After Victor Hansen).

71

Denmark; in Sweden found in most provinces north to Dlr. and Gstr.; in Norway only mentioned from the south (VAy, AK, Bø, Bv, Os, On); in Finland north to Oa, Sb and Kb. – North and Central Europe, especially towards the east rather abundant, ranging from the Netherlands and Belgium through Central Europe (south to Austria) to northern USSR, where it is widely distributed, eastwards to Siberia.

Biology. In stagnant fresh water, mainly in open, fairly eutrophic and usually somewhat clayey pools with some vegetation, often in deciduous woodland, though avoiding the more shaded sites. Also found in drift on the seashore. Most abundant in spring and autumn.

Note. Thomson (1853) mentioned that the species was widespread and also occurred in Lapland. This Lapland-record has been repeated in subsequent Swedish catalogues, until Lindroth (1960). However, compared to the known area of distribution, a Lappish occurrence is hardly reliable. As no Lappish specimens have been located, neither in Thomson's own collection nor in the Lapland collection of Zetterstedt (Nilsson, 1984), the Lappish record of the species must be deleted.

23. *Limnebius nitidus* (Marsham, 1802)
    Figs 110, 111, 114.

Hydrophilus nitidus Marsham, 1802, Ent. Brit. 1: 407 (sp. no. 15).
Hydrophilus picinus Marsham, 1802, Ent. Brit. 1: 407 (sp. no. 17).

1.4-1.7 mm. Black; pronotal and elytral margins reddish. Head and pronotum shining, with moderately dense and fine, very distinct punctuation; the reticulation strongly reduced on pronotum, only visible on lateral portions. Elytra somewhat shining, finely, but very distinctly reticulate, with fine (posteriorly very fine) punctuation. Appendages yellowish red, also femora pale.

113

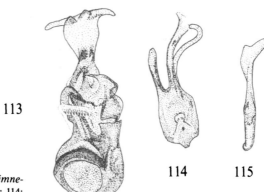

114   115

Figs 113-115. Male genitalia of *Limnebius*. – 113: *truncatulus* Thoms.; 114: *nitidus* (Marsh.); 115: *atomus* (Dft.).

72

♂: Pro- and mesotibiae somewhat enlarged distally. 6th visible abdominal sternite without long setae or other characteristics, but larger than in ♀ (see Figs 110, 111). Aedeagus, Fig. 114.

Distribution. Rare in Denmark, only known from a few localities in SJ, NEJ, NWZ and NEZ; in Sweden very rare, only found in Sk.; in Finland only recorded from N; not in Norway. – Widespread in West Europe: the British Isles, France and (?) Spain, eastwards to North Italy, Czechoslovakia and the Baltic States of USSR.

Biology. In clear, often rather eutrophic, running water, predominantly in wet mud at the edges of open, soft-bottomed streams, but also among vegetation in shallow water near the edges. Found numerously in July, when the newly emerged beetles appear, and also in autumn and spring.

24. *Limnebius aluta* Bedel, 1881
Pl. 1: 9.

Limnebius aluta Bedel, 1881, Faune Col. bass. Seine 1: 315.

1.1-1.3 mm. Black; pronotal and elytral margins reddish. Dorsal surface somewhat shining, hardly punctured but with distinct, though fine, reticulation. Anterior margin of labrum emarginate. Appendages reddish brown, femora darker, antennae yellow.

♂: Pro- and mesotibiae somewhat stronger enlarged distally than in ♀. 6th abdominal sternite almost as in *nitidus*.

Distribution. Widespread in Denmark, in the southern and eastern provinces rather common; in Sweden north to Vg., Nrk. and Gstr., in the south fairly common; southernmost Norway (Ø, AK, Bø); in Finland north to Tb. – North and Central Europe; ranges from SE. England and NE. France to Poland and Czecholovakia, from Fennoscandia to Austria and Italy.

Biology. In stagnant fresh water, especially at the edges of comparatively open, eutrophic pools with abundant vegetation, mostly among wet mos or in wet mud; particularly in deciduous woodland. April-June and again later in summer and autumn.

Subgenus *Bilimneus* Rey, 1883

Bilimneus Rey, 1883, Revue Ent. 2: 88.
Type species: *Hydrophilus atomus* Duftschmid, 1805, by present designation.
Bolimnius Rey, 1885a (emend.).
Bilimnius auctt.

Elytra (in the species treated here) with a distinct, though very fine sutural stria posteriorly. Pygidial hairtufts (in both sexes) so close-set that they appear as a single median tuft. Very small species. Male genitalia without parameres, rather simply tubular (Fig. 115).

Rey originally included two species in *Bilimneus,* but neither he nor anyone else have designated a type-species for the subgenus.

Limnebius atomus (Duftschmid, 1805)
Fig. 115.

Hydrophilus atomus Duftschmid, 1805, Fauna Austr. 1: 245.
Limnebius picinus auctt.; *nec* (Marsham, 1802).

1.0-1.2 mm. Black; pronotum and elytra brown, the latter darker in middle. Dorsal surface rather shining; head and pronotum smooth, without reticulation, and only very indistinctly punctured. Elytra with very fine reticulation and very fine punctuation. Anterior margin of labrum not emarginate. Elytra rather strongly narrowed posteriorly, with a very fine sutural stria posteriorly. Appendages yellowish brown, legs and maxillary palpi (particularly terminal segment) often darker.

♂: Pro- and mesotibiae slightly more enlarged distally than in ♀. Aedeagus, Fig. 115.

Distribution. So far not recorded from Scandinavia. – South and Central European species, becoming less frequent towards the north; in the North German lowland very rare and sporadic, found near Hamburg. An occurrence in southernmost Scandinavia is not impossible.

Biology. In running or stagnant fresh water, often in more or less temporary waters, under stones or in sandy soil at the edge of the water. In North Germany it is found in a shallow pond on a meadow together with *L. aluta* (Lohse, *in litt.*). Adults are found in May.

Family Spercheidae

Ocelli absent. Scutellum distinct. Prosternum well developed, not concealed by procoxae; antennal excavations vaguely defined; procoxal cavities closed posteriorly (Fig. 116). Abdomen with 5 visible sternites. Antennae (Fig. 117) 7-segmented (but with minute 3rd segment, so they appear 6-segmented), 2nd and 4th segments pubescent, thus similar to the three distal (pubescent) club-segments. Tarsi 5-segmented, basal segment minute, sometimes very indistinct; claw-segment as long as the preceeding segment. All trochanters distinct. Venation of hind wings of the cantharoid type (as Fig. 22). Aedeagus of the trilobed type (Fig. 10).

A rather small family, comprising only one genus, represented in all major zoogeographical regions.

Genus *Spercheus* Illiger, 1798

Spercheus Illiger, 1798, Verz. Käf. Preuss.: 241.
Type species: *Dytiscus emarginatus* Schaller, 1783, by monotypy.

Body contour interrupted between pronotum and elytra. Head rather wide, with large, more or less protruding eyes; clypeus large, its margins extended and distinctly bent upwards, anterior margin concave, often distinctly angularly emarginate in middle; labrum concealed under clypeus (dorsal view). Pronotum short and wide, with gently rounded sides, narrowed both anteriorly and posteriorly, middle portion (particularly in ♀) strongly convex, lateral portions more flattened. Scutellum fairly large, longer than wide. Elytra highly arched, extensively punctured, the punctures arranged in close-set, more or less regular rows; elytra in some exotic species with longitudinal ridges or raised tubercles. Ventral surface of body dull, shagreened and with dense hydrofuge pubescence. Legs moderately long, meso- and metatibiae distinctly curved.

The genus comprises only about 20 known species, occurring mainly in warmer climates. A single species occur in Europe.

The genus is distinctive in that the females carry the eggs in a bag on the ventral surface of the abdomen.

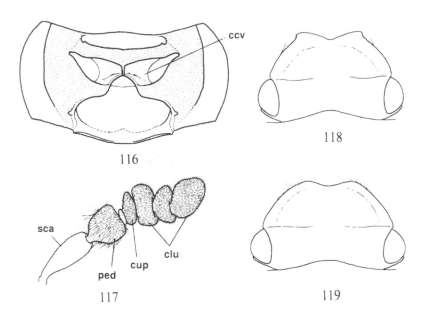

Figs 116-119. *Spercheus emarginatus* (Schall.). – 116: prosternum (legs removed); 117: right antenna; 118: male head; 119: female head.

75

25. *Spercheus emarginatus* (Schaller, 1783)
Figs 116-120; pl. 2: 1.

Dytiscus emarginatus Schaller, 1783, Schrift. naturf. Ges. Halle 1: 327.

5.5-7.0 mm. Piceous to black; clypeal and pronotal margins reddish; elytra brown or yellowish brown with small black spots, that medially become more close-set and mutually fused. Head and pronotum closely and moderately strongly punctured, in part often feebly and irregularly rugulose, especially on head. Elytra widest in or anterior to middle, with numerous, very close-set and somewhat irregular longitudinal series of dense and moderately strong punctures, and with rather indistinct and rudimentary longitudinal ridges. Maxillary palpi yellowish red, terminal segment black apically; antennae blackish, paler at base; legs dark reddish brown.

♂: Anterior angles of clypeus distinctly angular (Fig. 118).

♀: Anterior angles of clypeus rounded (Fig. 119).

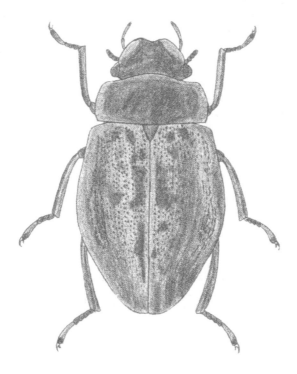

Fig. 120. *Spercheus emarginatus* (Schall.), length 5.5-7.0 mm. (After Victor Hansen).

76

Distribution. A widespread, but comparatively uncommon species; in Denmark recorded from most provinces; in Sweden scattered records north to Vstm. and Gstr.; not in Norway and Finland. – Europe; ranges from SE. England and NE. France through Central and East Europe to Siberia, from Scandinavia to North Italy and (?) the northern Balkans.

Biology. In stagnant, muddy or swampy, eutrophic fresh water with rich vegetation, usually at the roots of water plants at a depth of 20-50 cm, but also crawling freely among vegetation in the water. Adults are found mainly in the summer or in spring, where females carrying the characteristic egg-sac can be seen. Due to the rapid development of the larvae, full grown larvae are often found together with egg-carrying females. Numerous newly emerged adults are found in the middle of June. The species is not a very active flyer, but is occasionally (in May) taken in drift on the seashore.

Family Hydrochidae

Ocelli absent. Scutellum distinct. Prosternum well developed, not concealed by procoxae; antennal excavations vaguely defined; procoxal cavities markedly closed posteriorly (Fig. 121). Abdomen with 5 visible sternites, 5th sternite with a rather stiff, semitransparent lobe posteriorly (Figs 128, 129). Antennae (Fig. 122) 7-segmented with a pubescent 3-segmented club. Tarsi 5-segmented, basal segment minute, sometimes indistinct; claw segment as long as the preceeding segments. All trochanters distinct. Venation of hind wings of the cantharoid type (as Fig. 22). Aedeagus of the trilobed type (Fig. 10), often asymmetrical.

A rather small family, comprising only one genus, represented in all the major zoogeographical regions.

Genus *Hydrochus* Leach, 1817

Hydrochus Leach, 1817, Zool. Misc. 3: 90.
    Type species: *Silpha elongata* Schaller, 1783, by subsequent designation (Hope, 1838).

Body contour interrupted between pronotum and elytra. Head about as wide as pronotum, with large protruding globular eyes, and with a more or less distinct transverse v-shaped frontoclypeal furrow, that continues posteriorly in a narrow, often short and rather indistinct longitudinal furrow. Pronotum about as wide as long, with rather straight, or only slightly rounded (or sinuate) lateral margins; widest anterior to middle, hardly narrowed anteriorly, distinctly narrowed posteriorly; pronotum rather uneven, with large, though rather shallow impressions, arranged in two transverse rows, one in middle consisting of 3 impressions, and one behind this, consisting of 4 impressions. Head and pronotum punctured, in some species with distinct, though extremely shallow granulation between the punctures. Scutellum very small, elongate. Elytra usually rather long and narrow, with 10 strongly punctured striae; interstices very narrow and often (typically the alternate interstices) more or less markedly ridged, or even provided with raised tubercles. Ventral surface of body dull, almost velvety, with very fine and dense hydrofuge pubescence, quite uneven, with numerous pit-like depressions and large deep punctures, obtuse ridges, furrows etc.; abdominal sternites each with a strongly raised transverse bulge anteriorly, ranging across whole width (depressed posteriorly). Maxillary palpi moderately long, terminal segment slightly longer and thicker than penultimate. Legs moderately long and slender.

The genus comprises almost 100 known species; 4 of these occur in Fennoscandia and Denmark.

Key to species of *Hydrochus*

1   4th elytral interstice (in addition to 3rd and 5th) markedly
       raised behind middle (Fig. 123). Larger, 3.3-4.2 mm 2
–   4th elytral interstice not raised. Smaller, 2.4-3.2 mm 3
2 (1) Elytra about twice as long as wide, subparallel in anterior
       2/3; 4th interstice (besides posterior ridge) with a short
       distinct sub-basal ridge (Fig. 123). ♂: Aedeagus, Fig. 132.
       ♀: 5th visible abdominal sternite, Fig. 129 .... 26. *elongatus* (Schaller)
–   Elytra shorter, only about 3/4 longer than wide, from hu-
       meral angles distinctly widening posteriorly; 4th elytral in-
       terstice without distinct sub-basal ridge. ♂: Aedeagus, Fig.
       133. ♀: 5th visible abdominal sternite, Fig. 128 .. 27. *ignicollis* Motschulsky
3 (1) Elytra about 3/4 longer than wide, the striae subapically
       interrupted by a narrow transverse ridge, beyond which
       lie some enlarged, narrowly transverse punctures (almost
       as in Fig. 125) .... 28. *carinatus* Germar

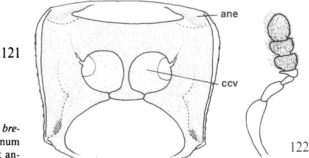

Figs 121, 122. *Hydrochus brevis* (Hbst.). – 121: prosternum (legs removed); 122: right antenna.

–   Elytra markedly broader, only about 1/2 longer than
     wide, without enlarged apical punctures and subapical
     transverse ridge, the striae continued almost to apex (Fig.
     126) .... 29. *brevis* (Herbst)

26. *Hydrochus elongatus* (Schaller, 1783)
    Figs 123, 124, 127, 129-132.

Silpha elongata Schaller, 1783, Schrift. naturf. Ges. Halle 1: 257.

3.5-4.2 mm. Piceous to black; head, pronotum and elytral ridges usually with a greenish iridescent lustre. Head and pronotum densely, and extremely shallowly granulate,
between the granules with fairly close-set and deep punctures. Elytra long and narrow,
about twice as long as wide, in anterior 2/3 subparallel, then rather strongly narrowed
apically, the striae very strongly and closely punctuated; interstices very narrow, 3rd
ridged about anterior third, 5th, 7th and 9th for almost entire length, and the 4th from
a little before middle to posterior quarter (Fig. 123); 4th interstice also with a short
sub-basal ridge which is as wide as, but slightly lower than, those of 3rd and 5th interstice. Ridges of alternate interstices anteriorly fused with a ridge on basal elytral margin, posteriorly fading. Elytral striae interrupted subapically by a rather bulging transverse ridge, beyond which lie some enlarged transverse apical punctures (Fig. 124).
Elytral ridging variable, yet always distinct. Appendages brown or reddish brown, an-

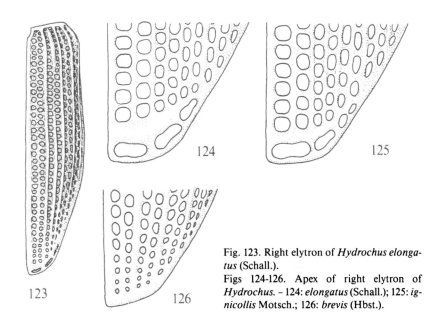

Fig. 123. Right elytron of *Hydrochus elongatus* (Schall.).
Figs 124-126. Apex of right elytron of
Hydrochus. – 124: *elongatus* (Schall.); 125: *ignicollis* Motsch.; 126: *brevis* (Hbst.).

tennal base paler, maxillary palpi (particularly at apex) and apex of claw segments darker.

♂: Aedeagus, Fig. 132.

♀: 5th visible abdominal sternite on each side narrowly excised (in ♂ simple) (Fig. 129).

Distribution. Widespread and not rare in Denmark; in Sweden only scattered records from the southern provinces, north to Upl.; not in Norway and Finland. – Europe; known from the British Isles, France, the Netherlands, Central Europe (widely distributed), eastwards to East Siberia, south to Italy (Toscana) and Yugoslavia. The total Palearctic area of distribution is still incompletely known due to confusion with *ignicollis,* and a re-examination of many older records is necessary.

Biology. In stagnant fresh water, particularly eutrophic, shallow clayey pools rich in vegetation, among which it lives; it is typical of fens in open country, but may also be found in deciduous woodland, though it avoids shaded sites. Mostly found in spring and autumn; eggs are laid in spring. Also, but not often, found in drift on the seashore.

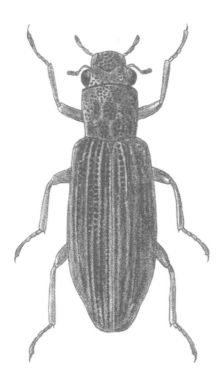

Fig. 127. *Hydrochus elongatus* (Schall.), length 3.5-4.2 mm. (After Victor Hansen).

27. *Hydrochus ignicollis* Motschulsky, 1860
    Figs 125, 128, 133.

Hydrochus ignicollis Motschulsky, 1860, Reis. Forsch. Amur-L. Schrenk 2 (2): 104.
Hydrochus elongatus auctt.; *nec* (Schaller, 1783).

3.3-4.0 mm. Very similar to *elongatus* in colour and size. Sculpturation of head and
pronotum as in *elongatus;* elytra slightly shorter, about 3/4 longer than wide, not subparallel, but distinctly widened from humeral angle to about posterior third, then
rather strongly narrowed apically; striae very strongly and closely punctured; ridging
of the interstices almost as in *elongatus,* except that 4th interstice is narrow and hardly
ridged sub-basally, and the transverse subapical ridge is usually much narrower and
hardly bulging (Fig. 125). Elytral ridges always well marked, less variable than in *elongatus.* Antennal club usually rather pale, not darker than base.
    ♂: Aedeagus, Fig. 133.
    ♀: 5th visible abdominal sternite not excised laterally (Fig. 128) (also simple in ♂).

    Distribution. Widespread and not rare in Denmark; in Sweden more frequent than
elongatus, found north to Ång.; in Norway in the southernmost provinces; widespread
in southern Finland, north to Oa and Kb. – The species has until recently been confused with *elongatus,* so its total Palearctic area of distribution is still not known in details. Besides Denmark and Fennoscandia it has been recorded from the British Isles,
France, the Netherlands, West and East Germany, Austria and the USSR (the Baltic
States and Ladoga); apparently not extending into Siberia.

    Biology. The habitat is similar to that of *elongatus,* but apparently *ignicollis*

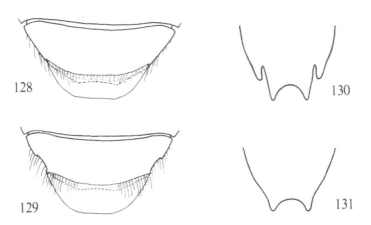

Figs 128, 129. Last abdominal sternite of *Hydrochus*-females. – 128: *ignicollis* Motsch.; 129:
elongatus (Schall.).
Figs 130, 131. 8th tergite of *Hydrochus elongatus* (Schall.). – 130: female; 131: male.

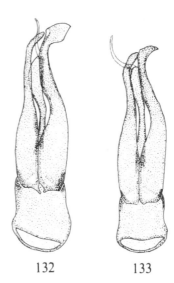

Figs 132, 133. Male genitalia of *Hydrochus*. – 132: *elongatus* (Schall.); 133: *ignicollis* Motsch.

132   133

prefers (? or tolerates) less eutrophic and more acid waters. Often found together with *brevis* and (rarely) *elongatus* and *carinatus*. Mainly in spring and autumn.

28. *Hydrochus carinatus* Germar, 1824

Elophorus crenatus Fabricius, 1792, Ent. Syst. 1 (1): 205 (nomen oblitum).
Hydrochus carinatus Germar, 1824, Ins. spec. nov.: 89.

2.4-3.0 mm. Black, without metallic reflections on head and pronotum ; elytral ridges usually with a faint bluish or greenish hue. Head and pronotum with fairly dense, strong and deeply impressed punctures, not granulate between the punctures. Elytral shape as in *ignicollis,* the striae very strongly and closely punctured; interstices very narrow, 3rd, 5th, 7th and 9th interstices raised above the others for almost their entire length, more or less markedly ridged, yet posteriorly more feebly so; 4th interstice not raised. Elytral striae interrupted subapically by a transverse, fairly narrow ridge; enlarged apical punctures narrowly transverse. Appendages reddish brown, base of antennae paler; maxillary palpi (especially at apex), femora and apex of claw segments darker.

Distribution. Common in Denmark; widespread in southern Sweden, north to Upl.; in Finland only a few findings from Ab and Ta; not in Norway. – Europe, ranging from Great Britain and France to northern Russia, from Fennoscandia to Central France, Corse, Sardinia, North Italy and the Balkans.

Biology. In stagnant, only occasionally running, fresh water, preferring shallow muddy or clayey waters with abundant vegetation, often at the same sites as *elongatus,* but more euryoecious. Occasionally taken in drift on the seashore. Mainly spring and autumn.

29. *Hydrochus brevis* (Herbst, 1793)
   Figs 121, 122, 126; pl. 2: 2.

Elophorus brevis Herbst, 1793, Natursyst. all. bek. in-u.ausl. Ins. 5: 141.

2.6-3.2 mm. Broader than the preceeding species. Black, without metallic reflections. Head and pronotum with fairly dense, strong and deep punctures, not granulate between the punctures. Elytra only 1/2 longer than wide, from humeral angle to about posterior third distinctly widening, then strongly narrowed apically; the striae very strongly punctured, punctures on the average larger than in the other species; interstices very narrow, 3rd, 5th, 7th and 9th interstices raised above the others and more or less markedly ridged, though posteriorly only feebly so; 4th interstice (as in *carinatus*) not raised. Elytra without subapical transverse ridge, and without enlarged apical punctures, the striae continued almost to apex (Fig. 126). Appendages dark reddish brown, antennae often paler (at least at base); maxillary palpi, femora and tarsi darker, sometimes almost blackish.

Distribution. Common in Denmark and most of Sweden north to Nb. and Lu.Lpm.; in Norway only mentioned from the southern parts; widespread in Finland north to ObS, less frequent towards the north. – Ranges from the British Isles and France to Siberia, from Fennoscandia to North Italy.

Biology. In stagnant water, predominantly in well established, eutrophic and well vegetated, shallow pools, often together with one or more of the other *Hydrochus*-spp. Sometimes in drift on the seashore. Mainly in spring and autumn.

Family Georissidae

Ocelli absent. Scutellum hardly visible. Procoxae fused with the trochanters, forming two large plates (Fig. 135), completely concealing the strongly reduced prosternum, that in middle forms only a simple, rather narrow chitinous bridge without coxal cavities (Fig. 134). Prosternum with very well defined antennal excavations. Abdomen with 5 visible sternites. Antennae (Fig. 136) 9-segmented with 3-segmented pubescent club. Tarsi 4-segmented; claw segment shorter than the preceeding segment. Hind wings often rudimentary, both in regard to size and venation, but distinctly of the cantharoid type (as Fig. 22). Aedeagus of the trilobed type (Fig. 10), with large basal piece.

A rather small family, comprising only one genus, represented in all major zoogeographical regions.

Genus *Georissus* Latreille, 1809

Georissus Latreille, 1809, Gen. Crust. 4: 377.
   Type species: *Pimelia pygmaea* Fabricius, 1801 (= *Byrrhus crenulatus* Rossi, 1794),
   by monotypy.
Georyssus auctt.

Body contour interrupted between pronotum and elytra. Head vertical, extensively granulate, the surface more or less uneven, often with impressions or obtuse ridges. Pronotum widest near base, strongly narrowed anteriorly, with only indistinct anterior angles; the lateral margins continued anteriorly into the strongly rounded margin of the anterior, extended and somewhat overhanging portion of pronotum, which to a great extent conceals the head. Pronotal surface variably granulate (at least partially), in the species treated here rather evenly convex; in other (mainly exotic) species rather uneven, with raised tubercles, obtuse ridges, depressions, etc. Scutellum minute, hardly visible. Elytra very short, as wide as long, strongly convex, normally with distinct humeral bulge; the lateral margins ventrally deflected inwards, closely clasping the body; elytral apex slightly extended. In the here treated species with 10 regular punctured elytral striae, and uniform feebly convex interstices; in some exotic species with alternate interstices raised above the others, being sometimes pronouncedly ridged and with reduced punctuation. Prosternum rudimentary, only weakly sclerotized; mesosternum short and wide, pentagonal, anteriorly more or less vertical; metasternum and abdominal sternites rather uneven, glabrous and somewhat shining; 1st abdominal sternite very large, rather convex, and strongly depressed posteriorly.

   The genus comprises a little more than 50 known species; only one is known from Fennoscandia and Denmark.

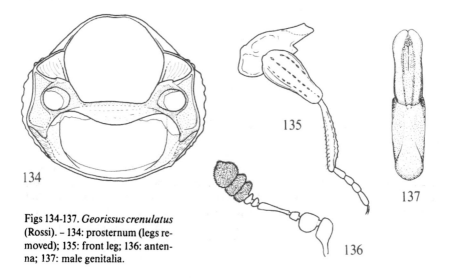

Figs 134-137. *Georissus crenulatus* (Rossi). – 134: prosternum (legs removed); 135: front leg; 136: antenna; 137: male genitalia.

30. *Georissus crenulatus* (Rossi, 1794)
  Figs 134-138; pl. 1: 12.

Byrrhus crenulatus Rossi, 1794, Mant. Ins. 2: 81.
Pimelia pygmaea Fabricius, 1801, Syst. Eleuth. 1: 133.

1.5-2.0 mm. Black, dorsal surface rather shining, with a very faint metallic hue. Head (apart from a smooth field on each side behind eyes) rather strongly and densely granulate. Pronotum wider than long, widest in posterior half, strongly narrowed anteriorly, posteriorly only slightly so; lateral margins feebly sinuate in front of middle. Pronotum anteromedially with a narrow longitudinal furrow and (on each side of this) in anterior third with a transverse furrow separating the anterior (overhanging and ex-

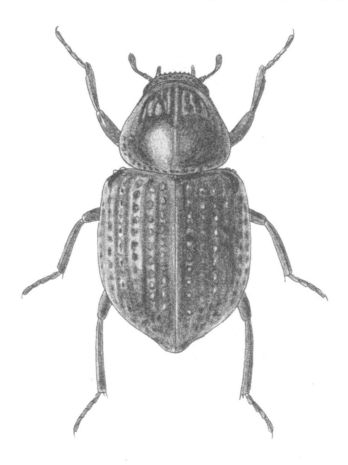

Fig. 138. *Georissus crenulatus* (Rossi), length 1.5-2.0 mm. (After Victor Hansen).

tensively sculptured) pronotal portion from the posterior (larger and smooth and shining) portion; the latter with a series of granules running along basal and lateral margins, and on each side (at some distance from lateral margin) with a shorter, but similar longitudinal series. Elytra strongly and regularly convex, and strongly but not very densely punctuated; interstices quite uniform, feebly convex, without distinct microsculpture. Appendages black, antennal base often brown.

Distribution. Widespread, but not particularly frequent; found in most parts of Denmark; in Sweden recorded from all provinces north to Dlr., and from Äng., Vb. and Nb.; in Norway only in southernmost parts (Ø, AK, AAy); in Finland only in the south (Al, Ab, N, St). – Widespread in Europe, from Fennoscandia to Central and South Europe, and from the British Isles to the Caucasus and Transbaikal.

Biology. At the edges of fresh water, often, but not necessarily at running water, where it lives on the banks in moist, sparsely covered clay or clayey sand. It is rather local, but often abundant, where it is, though it may easily be overlooked due to its rather sluggish behavior; furthermore, the beetle is normally covered by a coat of some clayish substance. Mainly found in May-June.

Family Hydrophilidae

Ocelli absent. Scutellum distinct. Prosternum more or less well developed, not concealed by procoxae, with or without distinct antennal excavations; procoxal cavities open posteriorly (almost as in Fig. 43). Abdomen with 5 visible sternites; seldom with a 6th (retractable) sternite visible; in a few exotic genera the number of visible abdominal sternites is reduced to 4. Antennae 7- to 9-segmented, with a pubescent 3-segmented club. Tarsi 5-segmented; only very seldom 4-segmented (meso- and metatarsi in *Cymbiodyta* and protarsi in *Berosus*-♂); claw segment almost always much shorter than preceeding segments. All trochanters distinct. Hind wing venation of the cantharoid type (Fig. 22). Aedeagus of the trilobed type (Fig. 10).

A large family, represented in all parts of the world and consisting of numerous genera, and about 2200 known species.

Key to subfamilies of Hydrophilidae

1 Pronotum with conspicuous longitudinal furrows (Fig. 139). Body contour interrupted between pronotum and elytra .... Helophorinae (p. 87)
– Pronotum without longitudinal furrows. Body contour normally evenly curved, only occasionally feebly interrupted between pronotum and elytra................................. 2

2 (1) First segment of meso- and metatarsi markedly longer

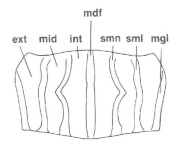

mdf

ext mid int smn sml mgl

Fig. 139. Furrows and intervals of pronotum of *Helophorus* (int = internal interval; mid = middle interval; ext = external interval; mdf = median furrow; smn = submedian furrow; sml = submarginal furrow; mgl = marginal furrow).

than 2nd, all tarsi 5-segmented. Maxillary palpi usually slightly shorter than antennae, and with 2nd segment strongly dilated . . . . Sphaeridiinae (p. 124)

– First segment of meso- and metatarsi markedly shorter than 2nd (except in *Cymbiodyta* which has 4-segmented meso- and metatarsi). Maxillary palpi normally as long as, or longer than, antennae, their 2nd segment usually not dilated 3

3 (2) Large or very large species, 14-48 mm. Meso- and metasternum markedly raised medially to form a common sternal keel . . . . Hydrophilinae (p. 211)

– Much smaller species, size not exceeding 10 mm. Meso- and metasternum often markedly raised medially, but not forming a common sternal keel 4

4 (3) Meso- and metatibiae with conspicuous fringes of long swimming hairs (Fig. 340). Scutellum distinctly longer than wide . . . . Berosinae (p. 218)

– Tibiae without fringes of long swimming hairs. Scutellum about as long as wide 5

5 (4) First two abdominal sternites covered by a bilobed hyaline mass supported by a fringe of long stiff yellowish setae (Fig. 326). Very small, semiglobular species, 1.3-1.7 mm, with antennal club rather compact (Fig. 327) . . . . . . Chaetarthriinae (p. 209)

– Abdomen without such hyaline mass and without long setae. Length usually well over 2.0 mm. Antennal club loose (Figs 281, 303, 304) . . . . Hydrobiinae (p. 165)

SUBFAMILY HELOPHORINAE

Body shape rather elongate, contour interrupted between pronotum and elytra. Head and pronotum with a distinctive pattern of impressed furrows. Meso- and metasternum not very strongly raised medially. Abdomen with 5 visible sternites. Maxillary palpi about as long as antennae, their 2nd segment not dilated; terminal segment longer than penultimate. Antennae 8- or 9-segmented, the club loose. Legs generally rather slender, tarsi 5-segmented; basal segment of meso- and metatarsi minute; dorsal face of meso- and metatarsi with fine, sometimes even rather long swimming hairs, or with small stiff setae. Tibiae without long swimming hairs.

The subfamily comprises only a single genus.

Genus *Helophorus* Fabricius, 1775

Elophorus Fabricius, 1775, Syst. Ent.: 66.
Helophorus Illiger, 1801 (emend.).
   Type species: *Elophorus minutus* Fabricius, 1775, proposed by Angus (1970d) (see below).

Head with more or less convex, but hardly protruding eyes, just behind these abruptly narrowed. Head and pronotum with granulate sculpture, that may be partially reduced, usually with metallic reflections. Head (Figs 161-165) with a transverse V-shaped frontoclypeal furrow, which continues posteriorly in a longitudinal frontal furrow. The frontal furrow is in some species narrow linear, in others rather wide and often expanding anteriorly. Pronotum (Fig. 139) with 5 longitudinal furrows separated by raised intervals, and at lateral margins with an extra furrow outside which is normally a raised border (the nomenclature of pronotal furrows and intervals used in the descriptions is given in Fig. 139). Pronotum arched transversely, in some species also longitudinally. Scutellum rather small. Elytra with 10 complete punctured striae, and in some species with an extra short intercalary stria at base between 1st and 2nd stria. The 11th (outermost) interstice strongly and sharply ridged (except in some exotic species), ventral portion of this ridge (the pseudepipleuron) often visible from below, outside the true epipleuron (Figs 140, 141). Elytra in most species yellowish to brown, with a distinctive pattern of black spots, of which the most characteristic is an arrowhead-shaped mark (the arrow-mark) across the suture in about middle; elytra only very seldom entirely black. Ventral surface dull, shagreened and with dense hydrofuge pubescence.

In general, the *Helophorus*-spp. are very uniform, and more of the species may be very difficult to determine. They show some variation in external characters, and occasionally aberrant individuals are found, deviating in characters that are normally good distinguishing characters (eg. pronotal granulation, shape of frontal furrow, colour). So the specimens must be examined very carefully, and comparison of the descriptions of closely related species is often necessary. The most reliable and constant characters are found in the male genitalia which ought to be extracted and placed in a drop of mountant, such as euparal.

As to the type-species of the genus there has been some confusion. Fabricius originally included two species in his genus, viz. *Silpha aquatica* Linnaeus and *Elophorus minutus* Fabricius. Strictly speaking, the type-species is *aquaticus,* as designated by

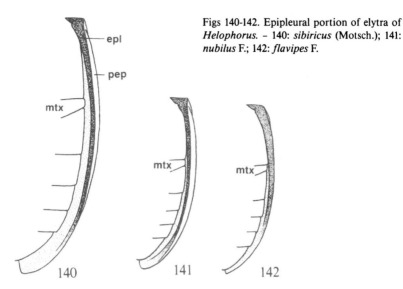

Figs 140-142. Epipleural portion of elytra of *Helophorus*. – 140: *sibiricus* (Motsch.); 141: *nubilus* F.; 142: *flavipes* F.

Latreille (1810). However, unfortunately this was overlooked by Kuwert (1886), who divided the genus into more subgenera, and placed *minutus* in *Helophorus* s.str. and *aquaticus* in a new subgenus, *Megahelophorus*. Kuwert's interpretation was followed by Sharp (1915-1916), who made a further division of the genus, and designated type-species for the respective subgenera (genera of Sharp), which Kuwert had not done. As type-species of *Helophorus* (s.str.) Sharp (1915a) chose *Elophorus griseus* Herbst, a species not originally included in the genus (but one year later, by Sharp, placed in synonymy with *minutus;* subsequently this synonymy was shown to be erroneous).

Throughout this century Kuwert's (and Sharp's) interpretation has been generally followed, and a change of the subgeneric names would not be expedient. This nomenclatoric problem is also pointed out by Angus (1970d), who is going to apply to the International Commission on Zoological Nomenclature, to reject Latreille's designation of *Silpha aquatica* L. as the type-species and replace it with *Elophorus minutus* Fabr., and to validate Illiger's emendation of the generic name as well as emendations of some of Kuwert's subgeneric names. Thus, though Angus' proposals have not been settled and may be rejected, I have followed the general interpretation rather than reintroducing names, which I hope need not to be revived.

The genus comprises almost 200 known species, represented in the Palearctic and the Nearctic regions; 30 species occur in Fennoscandia and Denmark or the adjacent areas.

Key to species of *Helophorus*

1 Intercalary stria present 2
– Intercalary stria absent 10

89

2 (1) Elytra entirely black, with raised tubercles on 3rd, 5th
and 7th interstice . . . . 32. *tuberculatus* Gyllenhal
– Elytra brown or yellowish, without tubercles 3
3 (2) Alternate interstices of elytra markedly raised above the
others. Pseudepipleura opposite metacoxae as wide as,
or wider than epipleura (Figs 140, 141) 4
– Alternate interstices of elytra at most weakly raised
above the others. Pseudepipleura opposite metacoxae
narrower than epipleura 7
4 (3) Terminal segment of maxillary palpi asymmetrical, with
straighter inner face (Fig. 147). Alternate interstices of
elytra more moderately ridged. Pseudepipleura opposite
metacoxae about as wide as epipleura (Fig. 140) . . . .
. . . . 33. *sibiricus* (Motschulsky)
– Terminal segment of maxillary palpi symmetrical (Fig.
148). Alternate interstices of elytra strongly ridged.
Pseudepipleura opposite metacoxae about twice as wide
as epipleura (Fig. 141) 5
5 (4) Smaller, 3.0-4.0 mm. Elytra at least 1/2 longer than
wide. Internal intervals of pronotum rather weakly nar-
rowed behind middle (Fig. 150) . . . . 31. *nubilus* Fabricius
– Larger, 4.0-5.6 mm. Elytra only about 2/5 longer than
wide. Internal intervals of pronotum strongly narrowed
behind middle (Figs 143, 144) 6
6 (5) Sides of pronotum distinctly sinuate posteriorly. Elytra
with sharp humeral angle (Fig. 144) . . . . *rufipes* (Bosc)
– Sides of pronotum evenly and rather feebly curved, not
sinuate posteriorly. Humeral angle rounded (Fig. 143) . . . . *porculus* Bedel
7 (3) Head and pronotum strongly and evenly granulate, the
granules very distinct and strikingly well defined, show-

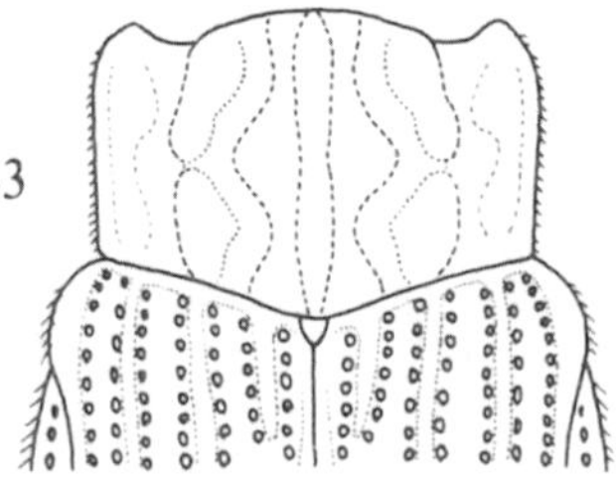

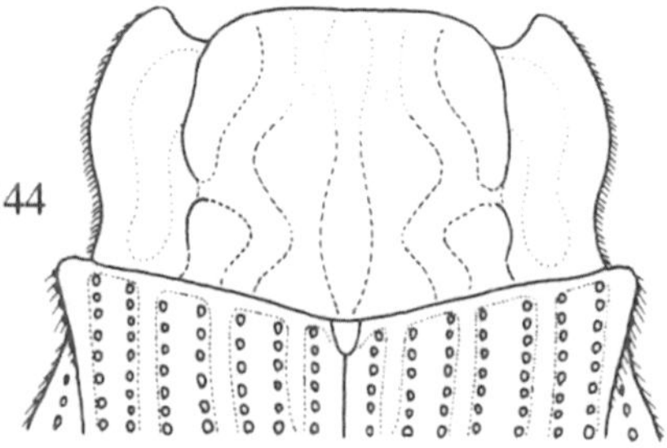

Figs 143, 144. Pronotum and anterior part of elytra of *Helophorus*. – 143: *porculus* Bedel; 144: *rufipes* (Bosc).

90

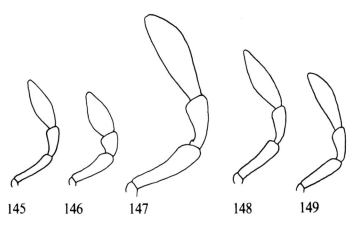

Figs 145-149. Maxillary palpus of *Helophorus*. – 145: *brevipalpis* Bedel; 146: *arvernicus* Muls.; 147: *sibiricus* (Motsch.); 148: *nubilus* F.; 149: *minutus* F.

ing no trace of fusing. Posterior margin of last visible abdominal sternite fairly strongly dentate (Fig. 166). . . . 36. *grandis* Illiger
– Head and pronotum more feebly granulate, the granules to some extent fusing and indistinct, especially on internal intervals of pronotum. Posterior margin of last visible abdominal sternite with finer (or indistinct) dention (Fig. 167) 8
8 (7) Frontal furrow narrowly linear (or sublinear), rather deep. Elytra about 1/2 longer than wide. Sides of pronotum normally markedly sinuate posteriorly . . . . 37. *strandi* Angus
– Frontal furrow almost always expanding anteriorly. Elytra narrower, normally about 3/5 longer than wide. Sides of pronotum normally more weakly sinuate 9
9 (8) Elytra dark brown, often with distinct lighter mottling. ♂: Parameres (dorsal view) making up less than half the total aedeagal length (index parameres: aedeagus = 0.42-0.47), their outer face normally distinctly rounded (Fig. 170). ♀: 9th horseshoe-shaped tergite (retracted in abdomen) with distinct lateral projections basal to which sides are distinctly concave (Fig. 173). . . . . 34. *aquaticus* (Linnaeus)
– Elytra often slightly paler, usually without distinct lighter mottling. ♂: Parameres (dorsal view) making up about half of the total aedeagal length (index parameres: aedeagus = 0.47-0.51), their outer face straighter, often *feebly sinuate* (Fig. 169). ♀: 9th tergite on the average a little narrower, with less distinct lateral projections, often

almost straight-sided (Fig. 172) .... 35. *aequalis* Thomson

10 (1) Legs and palpi very dark, often black, with distinct me-
        tallic reflections .... 40. *glacialis* Villa

–   Legs and palpi reddish or yellowish (sometimes brown),
        without any trace of metallic reflections 11

11 (10) Terminal segment of maxillary palpi symmetrical oval
        (Figs 145, 146)... 12

–   Terminal segment of maxillary palpi asymmetrical, with
        straighter inner face (Fig. 149) 13

12 (11) Pronotum moderately convex, lateral margins straight
        (or only weakly sinuate) in posterior half (Fig. 151) . 38. *brevipalpis* Bedel

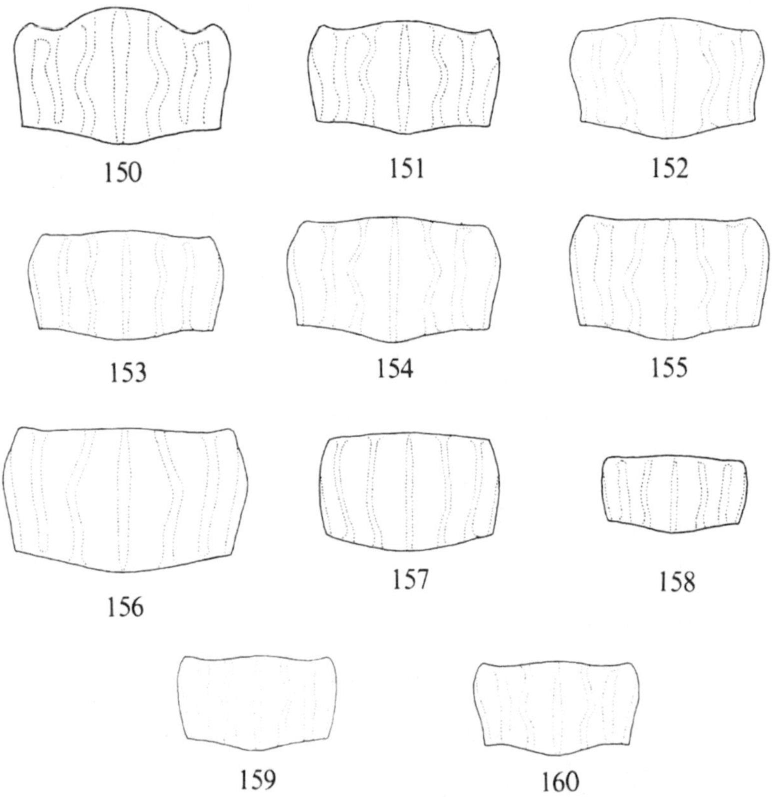

150

151

152

153

154

155

156

157

158

159

160

Figs 150-160. Pronotum of *Helophorus*. – 150: *nubilus* F.; 151: *brevipalpis* Bedel; 152: *arvernicus* Muls.; 153: *flavipes* F.; 154: *obscurus* Muls.; 155: *fulgidicollis* Motsch.; 156: *strigifrons* Thoms.; 157: *nanus* Sturm; 158: *granularis* (L.); 159: *minutus* F.; 160: *griseus* Hbst.

–   Pronotum strongly convex, lateral margins markedly si-
        nuate posteriorly (Fig. 152) . . . . 39. *arvernicus* Mulsant
13 (11) Antennae 8-segmented (Fig. 174). Frontal furrow nar-
        rowly linear (except in some rare specimens of *redten-*
        bacheri) 14
–   Antennae 9-segmented (Fig. 175). Frontal furrow nar-
        rowly linear or often wider, often expanded anteriorly 16
14 (13) Larger, 4.5-5.0 mm. Ground colour reddish brown or
        brown . . . . 50. *pallidus* Gebler
–   Smaller, 2.3-3.0 mm. Ground colour black, elytra paler 15
15 (14) Pronotum smooth and shining, often almost polished,
        normally finely punctured (almost without granulation).
        Head with a distinct pair of accessory furrows running
        (almost) parallel to median frontal furrow, between it
        and eyes (Fig. 163). . . . . 47. *nanus* Sturm
–   Pronotum rather coarsely granulate (though granules
        rather low). Head without such accessory furrows . . . .
        . . . . 48. *redtenbacheri* Kuwert
16 (13) Elytra with distinct (often very conspicuous) paler spots.
        Pseudepipleura broadly visible from below, opposite
        metacoxae about equally wide as epipleura (as Fig. 140).
        Frontal furrow broadly linear (Fig. 165) or expanded
        anteriorly (as Fig. 162) 17
–   Elytra normally without distinct (or at most vaguely de-

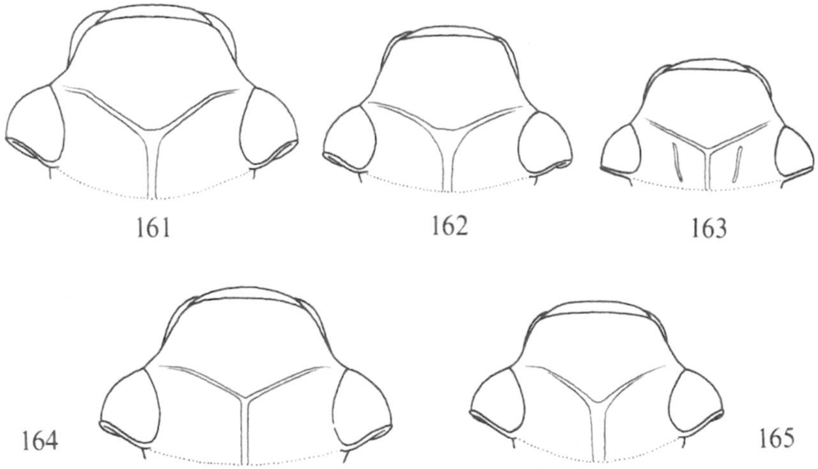

Figs 161-165. Head of *Helophorus.* – 161: *asperatus* Rey; 162: *flavipes* F.; 163: *nanus* Sturm; 164: *strigifrons* Thoms.; 165: *croaticus* Kuw.

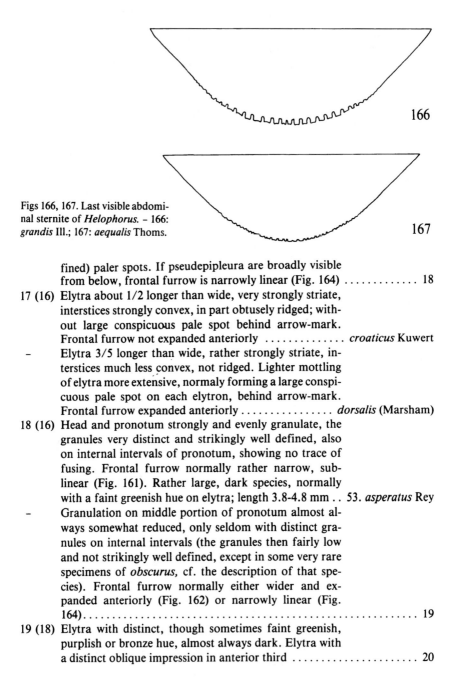

Figs 166, 167. Last visible abdominal sternite of *Helophorus*. – 166: *grandis* Ill.; 167: *aequalis* Thoms.

fined) paler spots. If pseudepipleura are broadly visible from below, frontal furrow is narrowly linear (Fig. 164) 18

17 (16) Elytra about 1/2 longer than wide, very strongly striate, interstices strongly convex, in part obtusely ridged; without large conspicuous pale spot behind arrow-mark. Frontal furrow not expanded anteriorly .... *croaticus* Kuwert

– Elytra 3/5 longer than wide, rather strongly striate, interstices much less convex, not ridged. Lighter mottling of elytra more extensive, normaly forming a large conspicuous pale spot on each elytron, behind arrow-mark. Frontal furrow expanded anteriorly .... *dorsalis* (Marsham)

18 (16) Head and pronotum strongly and evenly granulate, the granules very distinct and strikingly well defined, also on internal intervals of pronotum, showing no trace of fusing. Frontal furrow normally rather narrow, sublinear (Fig. 161). Rather large, dark species, normally with a faint greenish hue on elytra; length 3.8-4.8 mm .. 53. *asperatus* Rey

– Granulation on middle portion of pronotum almost always somewhat reduced, only seldom with distinct granules on internal intervals (the granules then fairly low and not strikingly well defined, except in some very rare specimens of *obscurus,* cf. the description of that species). Frontal furrow normally either wider and expanded anteriorly (Fig. 162) or narrowly linear (Fig. 164).. 19

19 (18) Elytra with distinct, though sometimes faint greenish, purplish or bronze hue, almost always dark. Elytra with a distinct oblique impression in anterior third 20

94

–   Elytra without any trace of metallic reflections, and with-
        out distinct impression in anterior third 21
20 (19) Submedian furrow of pronotum rather weakly curved
        outwards in middle, not distinctly angular (Fig. 153).
        Elytra normally tapered at apex. ♂: Aedeagus, Fig. 203
        .... 54. *flavipes* Fabricius
–   Submedian furrow of pronotum normally more strongly
        and markedly angularly curved outwards in middle (Fig.

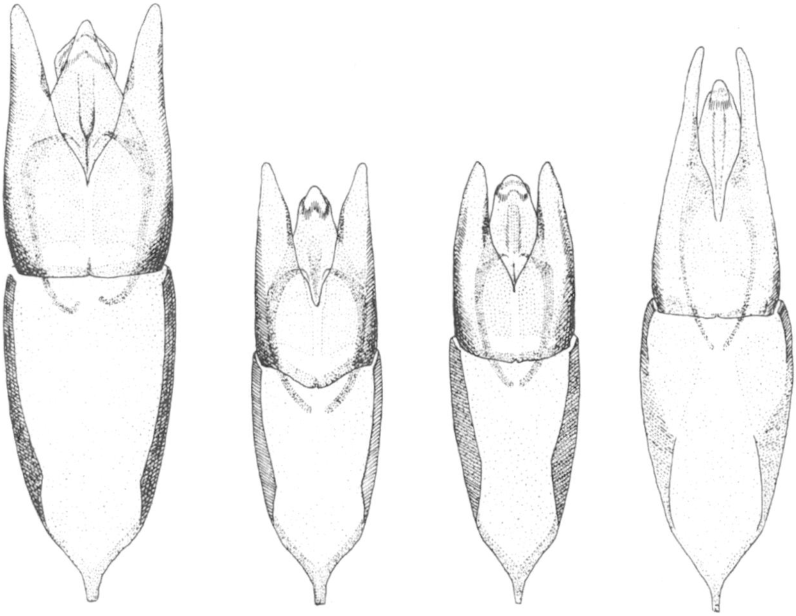

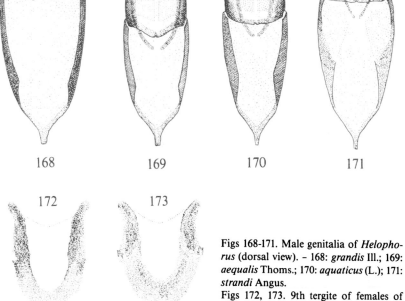

Figs 168-171. Male genitalia of *Helophorus* (dorsal view). – 168: *grandis* Ill.; 169:
aequalis Thoms.; 170: *aquaticus* (L.); 171:
strandi Angus.
Figs 172, 173. 9th tergite of females of
Helophorus (typical individuals). – 172:
aequalis Thoms.; 173: *aquaticus* (L.).

154). Elytra on the average a little wider, apex usually
more bluntly rounded. ♂: Aedeagus, Fig. 204 ..... 55. *obscurus* Mulsant

21 (19) Frontal furrow narrowly linear, rather deep (Fig. 164).
Pronotum strongly convex, also distinctly convex longitudinally, or internal intervals strongly convex about
middle, somewhat humpy 22

– Frontal furrow appearing more shallow, normally expanded anteriorly (as Fig. 162). Pronotum less strongly
convex, hardly convex longitudinally, internal intervals
not humpy 24

22 (21) Pseudepipleura narrow, not visible from below (as Fig.
142). Pronotum rather evenly convex, with uniform,
close and distinct, but rather feeble granulation; granules
normally visible also on internal intervals (though often
slightly weaker). Length, 3.0-4.0 mm .... 51. *laticollis* Thomson

– Pseudepipleura broadly visible from below, opposite
metacoxae about as wide as epipleura (as Fig. 140). Internal intervals of pronotum more or less humpy, normally with strongly reduced granulation, or even simply
punctured 23

23 (22) Larger, 3.5-4.2 mm. Pronotum (Fig. 156) widest just be-

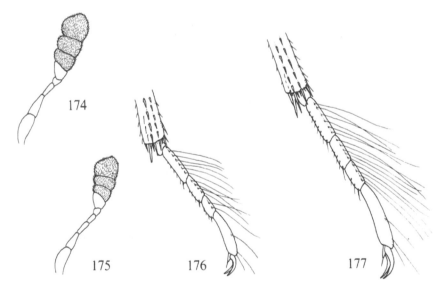

Figs 174, 175. Antennae of *Helophorus*. – 174: *nanus* Sturm; 175: *granularis* (L.).
Figs 176, 177. Metatarsi of *Helophorus*. – 176: *minutus* F.; 177: *fulgidicollis* Motsch.

fore middle, the sides a little straightened in posterior
half; strongly convex (though longitudinally only slightly
so), internal intervals usually markedly humpy .. 52. *strigifrons* Thomson
–   Smaller, 2.5-3.2 mm. Pronotum widest just behind middle, the sides evenly rounded (almost as in Fig. 157);
strongly arched and distinctly convex longitudinally . 49. *pumilio* Erichson
24 (21) Elytra yellowish brown, rather strongly striate. Pronotum strongly, rather evenly granulate; granules on internal intervals rather feeble, but usually distinct. Metatarsal claw segment (excluding claws) distinctly longer
than 1st + 2nd segment (Fig. 177). Length 3.2-4.5 mm .
        .... 45. *fulgidicollis* Motschulsky
–   Elytra paler, striae finer (except *granularis:* 2.2-2.8 mm).
Pronotum usually more weakly granulate, especially on
internal intervals (except *granularis* and some *lapponicus).* Metatarsal claw segment (excluding claws) not lon-
ger than 1st + 2nd segment (Fig. 176) (except *lapponicus*) 25
25 (24) Granulation of pronotum very reduced, at most visible
on external intervals (and only weakly so). Pronotum
very shining, bright metallic. Elytra very pale, finely
striate. ♂: Aedeagus characteristic (Fig. 191) ...... *longitarsis* Wollaston
–   Granulation of pronotum more distinct, also distinct on
middle intervals 26
26 (25) Pronotum rather strongly narrowed posteriorly, the sides
straightened posteriorly and normally feebly sinuate;
pronotal granulation rather fine, on internal intervals
usually reduced to simple punctuation. Elytra pale, finely
striate. ♂: Aedeagus characteristic (Fig. 190) .... 46. *griseus* Herbst
–   Pronotum a little less narrowed posteriorly, the sides
more evenly rounded, usually also posteriorly (though
here often a little straighter) 27
27 (26) Metatarsal claw segment (excluding claws) distinctly longer than 1st + 2nd segment (as Fig. 177). Size on the
average larger, 3.3-4.0 mm. ♂: Aedeagus, Fig. 192 ....
        .... 44. *lapponicus* Thomson
–   Metatarsal claw segment (excluding claws) not longer
than 1st + 2nd segment (Fig. 176). Size on average smaller, 2.2-3.6 mm 28
28 (27) Pronotum (Fig. 158) rather strongly convex, with rather
even, close and usually coarse, though very low granulation. Marginal furrows narrow, hardly paler. Submedian furrow only rarely angled in middle. Small, 2.2-
2.8 mm. ♂: Aedeagus, Fig. 187 .... 41. *granularis* (Linnaeus)
–   Pronotum (Fig. 159) normally weaker convex, granula-

tion (especially on internal intervals) weaker. Marginal furrow wider and often paler; if marginal furrow is not distinctly paler, submedian furrow is usually distinctly angled in middle. On average larger, 2.8-3.6 mm 29

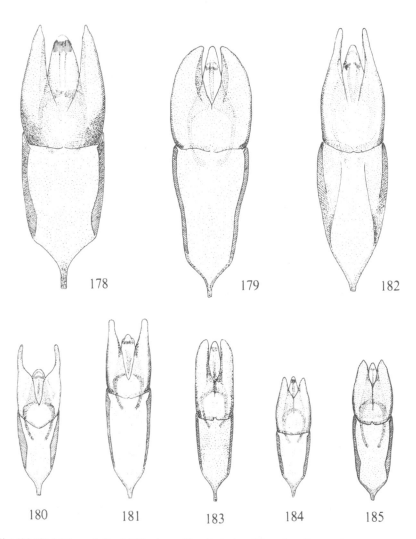

Figs 178-185. Male genitalia of *Helophorus* (dorsal view). – 178: *rufipes* (Bosc); 179: *porculus* Bedel; 180: *nubilus* F.; 181: *tuberculatus* Gyll.; 182: *sibiricus* (Motsch.); 183: *glacialis* Villa; 184: *brevipalpis* Bedel; 185: *arvernicus* Muls.

29 (28) Pronotum a little more convex and rather strongly granulate, the margins reddish or brownish, not conspicuously paler. ♂: Aedeagus, Fig. 188 . . . . 42. *discrepans* Rey
- Pronotum a little less convex, normally more weakly granulate, anterior and lateral margins normally distinctly paler, usually broadly yellow. ♂: Aedeagus, Fig. 189 . .
. . . . 43. *minutus* Fabricius

Subgenus *Empleurus* Hope, 1838

Empleurus Hope, 1838, Col. Man. 2: 149.
Type-species: *Elophorus nubilus* Fabricius, 1776, by original designation.

Elytra with intercalary stria. Interstices 1, 3, 5, 7, 9 and 11 strongly ridged; pseudepipleura opposite metacoxae about twice as wide as epipleura. Suprapleural portion of pronotum evenly wide throughout. Terminal segment of maxillary palpi symmetrical (Fig. 148). Tarsi on their dorsal face with small stiff setae (without long swimming hairs). Antennae 9-segmented.

Helophorus rufipes (Bosc, 1791)
Figs 144, 178.

Opatrum rufipes Bosc, 1791, Bull. Soc. Philomat. 1 (2): 8.

4.5-5.6 mm. A very broad species. Brown or reddish brown; elytra more yellowish brown, with distinct black sutural arrow-mark, a small black spot opposite this on interstice 7, and with some scattered brownish spots. Head and pronotum with distinct, rather uneven granulation; the granules to some extent with fine setae, in pronotal furrows however generally much smaller and without setae. Pronotum (Fig. 144) almost 3/4 wider than long, hardly narrowed posteriorly; lateral margins rounded anteriorly, markedly sinuate posteriorly, hind angles therefore sharply angled; anterior margin strongly bisinuate, anterior angles projecting; internal intervals anteriorly strongly depressed, superficial, in middle wide and strongly bulging, behind middle strongly narrowed, then again widening and bulging basally; middle intervals strongly bulging before and behind middle, just behind middle interrupted; lateral portions of pronotum (outside external furrows) impressed and broadly flattened, external intervals very shallow and indistinct. Elytra only about 2/5 longer than wide, fairly strongly striate; alternate (ridged) interstices each with a conspicuous series of close-set, small stiff setae; the other interstices with similar but less conspicuous series. Interstice 3 evenly ridged throughout, not depressed before middle. Elytra with sharply angled humeral angle (Fig. 144). Appendages yellowish red.
♂: Aedeagus, Fig. 178.

Distribution. So far not recorded from Scandinavia. – West European species; the

British Isles, France, southwards to the Iberian Peninsula and North Africa; in the Mediterranean particularly in the western parts, but also widespread in Italy; only scattered records from Yugoslavia and Greece. The species is so far not recorded from North Germany, but an occurrence in SW. Scandinavia is not impossible.

Biology. On sandy or clayey, sunny ground at roots of different plants, e.g. turnips, which may sometimes be damaged by the beetle, or perhaps, mainly its larva, which is probably phytophagous (as in *porculus*); it has, however, been suggested that it feeds on Halticine- and *Sitona*-larvae. Adults in spring and summer.

Helophorus porculus Bedel, 1881
   Figs 143, 179.

Helophorus porculus Bedel, 1881, Faune Col. bass. Seine 1: 298.

4.0-5.0 mm. Similar to *rufipes* in colour and shape. Granulation of head and pronotum almost as in *rufipes*. Internal intervals of pronotum anteriorly narrow, but not superficial, in middle strongly angularly widened, behind middle strongly narrowed, and then again strongly widened towards base (Fig. 143); middle and external intervals almost as in *rufipes*. Pronotum not quite 3/4 wider than long, almost parallel-sided, the sides evenly and only slightly rounded, not sinuate posteriorly; anterior margin strongly bisinuate. Elytra as in *rufipes,* except for humeral angle, which is rounded (Fig. 143).
   ♂: Aedeagus, Fig. 179.

Distribution. So far not recorded from Scandinavia. – Western Europe; the British Isles, France, southwards to Spain and North Africa, widespread in the Mediterranean (particularly in the west), eastwards to the northern Balkans and European Turkey; in NW. Europe eastwards to the Netherlands, where it is more frequent than *rufipes,* and West Germany (north to Holstein). Perhaps the species may also be found in SW. Scandinavia.

Biology. On sandy or clayey, sunny ground, at roots of plants or in rotting vegetables or the like, often on cultivated soils, where it may be a serious pest of turnips. The larva is phytophagous, sitting at the leaf bases, where it eats holes into the stem. Also the adult beetle feeds on the plants. It is clearly nocturnal, in the daytime being well hidden at the roots. Adults are found in spring and summer; eggs are laid in July-August. The larvae has a comparatively long lifetime, until following spring, when it goes down in the soil to pupate, and – after about six weeks – the new adults emerge. During winter the larva normally stays in the uppermost layer of the soil, but in cultivated soils it may go as far down as 30 cm.

31. *Helophorus nubilus* Fabricius, 1776
   Figs 141, 148, 150, 180; pl. 2: 3.

Elophorus nubilus Fabricius, 1776, Gen. Ins.: 213.

3.0-4.0 mm. Brown or reddish brown, lateral margins of pronotum paler. Elytra yellowish brown, with darker, usually black spots, somewhat mottled. Head and pronotum with distinct granulation, the granules with fine setae, except in the furrow where granules are very small. Pronotum about 1/2 wider than long, narrowed posteriorly; anterior margin almost as in the two preceeding species; the intervals more even and less bulging than in these species, internal intervals somewhat narrowed behind middle, but less strongly (Fig. 150). Elytra at least 1/2 longer than wide, fairly strongly striate; alternate (ridged) interstices each with a conspicuous series of close-set, small stiff setae; the other interstices with similar, but less conspicuous series. Ridge of interstice 3 somewhat depressed (but not interrupted) in anterior third. Humeral angle rounded. Antennae and maxillary palpi brownish, terminal segment of the latter darker; legs yellowish brown, tarsi slightly darker.

♂: Aedeagus, Fig. 180.

Distribution. Comparatively widespread and fairly common in southern Scandinavia; entire Denmark; in Sweden north to Dlr.; in Norway only Ø and AK; in Finland only in the south (Al, Ab, N, Sa). Ranges from West Europe to Siberia, from Fennoscandia to Spain, Central Italy and the Balkans.

Biology. On sandy or clayey, sunny ground, also on cultivated soil, usually rather dry places, but also on more humid habitats, e.g. near water, but only accidentally in water. It is found at the roots of various plants, in decaying debris, such as rotting vegetables, compost heaps etc. The larva has been recorded as a pest of wheat; life cycle probably as in *porculus*. Adults are taken from spring to autumn; they are not very active flyers, but are sometimes found in drift on the seashore. Very occasionally found in nests of different ants (e.g. *Lasius, Tetramorium*).

Subgenus *Cyphelophorus* Kuwert, 1886

Kyphohelophorus Kuwert, 1886b, Wien. ent. Ztg. 5: 223.
Cyphelophorus Kuwert; Seidlitz, 1888, Fauna Balt. (2): 115 (emend.).
    Type species: *Elophorus tuberculatus* Gyllenhal, 1808, by subsequent designation (Sharp, 1915a).

Elytra with intercalary stria. Interstices 3, 5 and 7 with raised tubercles, interstice 11 strongly and sharply ridged, pseudepipleura broadly visible from below, opposite metacoxae more than twice as wide as epipleura. Suprapleural portions of pronotum evenly wide throughout. Terminal segment of maxillary palpi symmetrical. Tarsi on their dorsal face with fine hairs. Antennae 9-segmented.

32. ***Helophorus tuberculatus* Gyllenhal, 1808**
    Fig. 181; pl. 2: 4.

Elophorus tuberculatus Gyllenhal, 1808, Ins. Suec. 1 (1): 129.

2.8-3.8 mm. An easily recognized species, characterized by its dark colour and elytral tubercles. Black, dorsal surface, especially elytra shining. Head and pronotum with a faint reddish bronze hue, strongly and closely granulate, with very distinct granules. Elytra finely striate, with an oblique depression in anterior third, and behind this with raised tubercles, normally 3 on 3rd interstice, 3 on 5th interstice and 2 on 7th interstice (often also with a weak tubercle on 9th interstice). Appendages very dark, almost blackish brown.

♂: Aedeagus, Fig. 181.

Distribution. Comparatively rare and local, but recorded from most provinces of Denmark, Sweden and Finland; in Norway only recorded from AK and TEy. – Widespread throughout northern Europe, from Great Britain and NE. France to Siberia, southwards to South Germany; also in North America.

Biology. A tyrphobiontic species, typically found in *Sphagnum*-bogs or at oligotrophic and acid pools on moors. Contrary to most other *Helophorus*-spp. (except subgen. *Empleurus*) it does not live in the water, but well hidden among wet mosses (mainly *Sphagnum*), where it is usually taken singly or only few in numbers. It is found from March to August, predominantly in spring, when it has an active period of flight, and may be found at random under very different conditions, often far from its natural habitat, e.g. regularly found in drift on the seashore. Hibernation probably in the adult stage.

Subgenus *Gephelophorus* Sharp, 1915

Gephelophorus Sharp, 1915c, Entomologist's mon. Mag. 51: 198.
Type species: *Helophorus auriculatus* Sharp, 1884, by original designation.

Elytra with intercalary stria. 11th interstice strongly ridged; pseudepipleura opposite metacoxae about as wide as epipleura (Fig. 140). Suprapleural portion of pronotum wide anteriorly, narrowed posteriorly. Terminal segment of maxillary palpi asymmetrical, the inner face being straighter than the outer face (Fig. 147). Tarsi with fine swimming hairs on their dorsal face. Antennae 9-segmented.

33. *Helophorus sibiricus* (Motschulsky, 1860)
Figs 140, 147, 182.

Empleurus sibiricus Motschulsky, 1860, Reis. Forsch. Amur-L. Schrenk 2 (2): 104.
Helophorus fennicus auctt.; *nec* Paykull, 1798.

4.7-5.8 mm. A rather variable, but easily recognizable species. Reddish brown, ventral face darker. Head and pronotum with metallic, usually reddish or greenish reflections; elytra pale brown with dark spots, including a distinct sutural arrow-mark. Head and pronotum closely and rather coarsely granulate, the granules with fine distinct setae. Frontal furrow deep and narrow, subparallel. Pronotum widest in about anterior third, strongly narrowed posteriorly; lateral margins slightly sinuate posteriorly. Elytra with interstices 1, 3, 5 and 7 markedly raised above the others (but less strongly ridged than in subgen. *Empleurus*); each interstice with a distinct series of close-set fine setae, which are equally distinct on raised and non-raised interstices. Appendages yellowish brown; apex of maxillary palpi, antennal club and apex of claw segments darker.

    ♂: Aedeagus, Fig. 182.

    Distribution. Widespread in northern Fennoscandia; in Sweden south to Jmt.; in Norway recorded from the northern parts (TRi, Fi, Fn, Fø) and a few provinces further south (HEn, On, STi); northern Finland, south to Ks.; not in Denmark. – A holarctic species, in the Palearctic region ranging from northern Fennoscandia through Russia and Siberia (where it is widely distributed) to Mongolia and China; in the Nearctic region mentioned from Alaska and Canada.

    Biology. In Scandinavia the species apparently prefers the edges of rivers, perhaps mainly sandy banks; but it is also (particularly in the more eastern parts of its range) taken in shallow grassy pools. Life cycle probably as in *aequalis,* the egg cocoons are placed in sand at the edge of the water. Adults in spring and summer, newly emerged specimens are found in August.

Subgenus *Meghelophorus* Kuwert, 1886

Megahelophorus Kuwert, 1886b, Wien. ent. Ztg. 5: 226.
Meghelophorus Kuwert; Sharp 1915a, Entomologist's mon. Mag. 51: 3 (emend.).
    Type species: *Silpha aquatica* Linnaeus, 1758, by subsequent designation (Sharp, 1915a).

Elytra with intercalary stria. 11th interstice sharply ridged; pseudepipleura not visible from below, or, if visible, narrower than epipleura. Terminal segment of maxillary palpi asymmetrical, with straighter inner face. Tarsi with fine swimming hairs on their dorsal face. Antennae 9-segmented.

34. *Helophorus aquaticus* (Linnaeus, 1758)
    Figs 170, 173.

Silpha aquatica Linnaeus, 1758, Syst. Nat. (10) 1: 362.
Helophorus aequalis auctt.; *nec* Thomson, 1868.

4.0-5.5 mm. Black or pitchy brown; head and pronotum metallic, greenish and bronze, mostly rather dark; pronotum usually greenish, laterally and in the furrows bronze or purplish, but sometimes almost uniformly purplish or bronze. Elytra brown, usually with a faint (mostly greenish) metallic hue, in middle with a black arrow-mark, laterally with a few additional black spots, often with distinct lighter mottling. Head and pronotum with close, but usually rather feeble and to some extent reduced granulation; the granules on head and internal intervals of pronotum somewhat fused and indistinct. Frontal furrow normally expanded anteriorly, only seldom linear. Lateral margins of pronotum slightly sinuate posteriorly. Elytra about 3/5 longer than wide, moderately striate; interstices 1, 3, 5 and 7 hardly raised above the others. Last visible abdominal sternite along posterior margin with fine, but distinct teeth (as Fig. 167). Appendages yellowish brown; apex of maxillary palpi, antennal club and claw segments (distally) darker.

♂: Aedeagus (Fig. 170) with parameres making up less than half of the total aedeagal length (index parameres: aedeagus, dorsal view! = 0.42-0.47). Outer face of parameres normally rounded, though often weakly so (see also note under *aequalis*).

♀: 9th tergite (retracted in abdomen) with distinct lateral projections, basal to which the sides are distinctly concave (Fig. 173) (see also note under *aequalis*).

Distribution. Rare and sporadic in eastern Denmark, in recent time only found in SJ and EJ; widespread in Finland; not found in Sweden and Norway. – Europe; from the Spanish highland through Central France to Denmark, eastwards to the Urals, from Fennoscandia to Italy (the Appennines), the Balkans and Asia Minor.

Biology. In stagnant fresh water, usually in small, shallow and muddy pools. It seems to be less euryoecious than *aequalis* and apparently prefers more shaded waters, often on boggy ground. Life cycle as in *aequalis;* egg cocoons (containing about 15 eggs) are placed in mud at the edge of the water. Adults mainly in spring and autumn.

Note. The species has until quite recently been considered conspecific with *aequalis* (see discussion under that species), and the name *aquaticus* has been the source of some confusion, since different authors have interpreted it differently, and in many cases applied it to either *grandis* or *aequalis*. Conversely, the name *aequalis* has often been applied to *aquaticus*. In consequence, almost all older records of these three species should be considered most unreliable. I have revised almost all available material from the museum collections of Denmark, Norway, Sweden and Finland, as well as a large number of specimens from private collections, and virtually all records included here are based on these specimens. It should be mentioned, that Lindroth (1960) gives some records from northern Norway and northern Finland, which I have not been able to confirm (for any of the three species). It is very likely that these records concern other species (probably *strandi,* in particular), and they are therefore omitted.

35. *Helophorus aequalis* Thomson, 1868
Figs 167, 169, 172; pl. 2: 7.

Helophorus aequalis Thomson, 1868, Skand. Col. 10: 300.
Helophorus aquaticus auctt.: nec (Linnaeus, 1758).

4.6-6.2 mm. Extremely closely related to *aquaticus,* but on the average a little larger, head and pronotum brighter metallic, and elytra usually without distinct lighter mottling. However these characters are very variable in both species, and there seems to be no reliable external characters separating the two.

♂: Aedeagus (Fig. 169) with parameres making up about half of the total aedeagal length (index parameres: aedeagus, dorsal view! = 0.47-0.51). Outer face of parameres straight or almost so, often feebly sinuate (see also note below).

♀: 9th tergite (retracted in abdomen) slightly narrower than in *aquaticus*, with less distinct lateral projections, often almost straight-sided (Fig. 172) (see also note below).

Distribution. Very common in Denmark, and in Sweden north to Vstm. (perhaps further north); widespread in Norway; in Finland only in the south (Al, Ab, N, Ta). – Europe, ranging from Fennoscandia to North Spain, Italy and the Balkans, from West Europe (including the British Isles) to western part of the USSR; in the south ranging further east, to Asia Minor and the Caucasus.

Biology. A fairly euryoecious species, preferring stagnant fresh water with grassy bottom; usually in open, rather eutrophic, shallow pools, but also in the slower reaches of streams or (?accidentally) in saline waters. The eggs are normally laid in spring; the egg cocoons, containing about 12 eggs each, are placed in mud at the edge of the water. Sometimes are the eggs laid already in late autumn, but in any case the larvae do not emerge until the following spring (or early summer); larvae full grown in about two or three weeks. Adults are mainly found in spring and autumn; they are sometimes found in drift on the seashore.

Note. *H. aequalis* was until recently considered conspecificic with *aquaticus*, but is separated and considered a distinct species by Angus (1982) who, on the basis of chromosome analysis showed a number of differences between the two forms. The aedeagal characters and the shape of 9th tergite in females (see description) vary in both forms, and they may be very difficult to recognize, even on these characters (especially some of the specimens from Finland). However, Angus has examined the karyotypes (which clearly separate the two forms) and the male genitalia in a number of specimens, and found that aedeagal variation is produced without hybridization. Further the egg cocoons and the third instar larvae of *aquaticus* and *aequalis* seem to be separable, so they are consequently here considered as good species. The modern distribution and fossil records also seem to support this view. (See also note under *aquaticus*).

36. *Helophorus grandis* Illiger, 1798
   Figs 166, 168.

Elophorus grandis Illiger, 1798, Verz. Käf. Preuss.: 272.
Helophorus aquaticus auctt.; *nec* (Linnaeus, 1758).

6.0-8.0 mm. Similar to *aquaticus* and *aequalis*, but on the average larger; metallic reflections of head and pronotum somewhat dull, greenish, pronotum laterally and in the furrows purplish; the granulation stronger than in the two preceeding species, the granules very distinct and strikingly well defined, not fusing even on the internal intervals of pronotum. Elytra with a stronger, oblique impression in anterior third, and with interstices 1, 3, 5 and 7 distinctly raised above the others. Dentition of posterior margin of last visible abdominal sternite stronger than in the other species of the subgenus (Fig. 166).

♂: Aedeagus, Fig. 168.

Distribution. In Denmark widespread, and common in the eastern parts; in Sweden widespread north to Nb. (not known from Lapland); in Norway only in the south (Ø, AK, Bø, VE); in Finland only a few specimens from Al and Ka. – Ranges from West Europe to the Urals, from Fennoscandia to North Spain, the Alps and the Carpathians; also in North America (Eastern USA). – Some closely related species, that were earlier considered subspecies of *grandis,* occur in the Mediterranean area.

Biology. In stagnant fresh water, preferring eutrophic, more or less open and often temporary pools with clayey and grassy bottom. Life cycle as in *aequalis,* egg cocoons placed in mud at the edge of the water. Mainly in spring and autumn; sometimes found in drift on the seashore.

37. *Helophorus strandi* Angus, 1970
   Fig. 171.

Helophorus strandi Angus, 1970d, Acta zool. fenn. 129: 46.
Helophorus bergrothi auctt.; *nec* J. Sahlberg, 1880.

4.5-5.0 mm. Black or pitchy brown, marginal furrows of pronotum brown or slightly reddish; head and pronotum blackish bronze, sometimes with reddish reflections; elytra dark brown, sometimes reddish, with black arrow-mark and a black spot opposite this on 7th interstice. Head and pronotum closely and coarsely granulate, the granules to some extent fusing, especially on internal intervals of pronotum. Frontal furrow narrowly linear, occasionally with a suggestion of expansion anteriorly. Pronotum widest in anterior third or fourth, distinctly sinuate posteriorly (normally stronger than in the preceeding species); the furrows narrow, rather shallow. Elytra rather wide, about 1/2 longer than wide, strongly striate, interstices convex, 1st, 3rd, 5th and 7th interstices raised a little above the others (especially in large specimens). Posterior margin of last visible abdominal sternite with very fine or indistinct dentition. Appendages mid or dark brown; apex of maxillary palpi at most slightly darker, antennal club always dark brown, claw segments gradually darkened distally.

♂: Aedeagus, Fig. 171.

Distribution. A boreal species, restricted to a few provinces in northern Fennoscandia; in Sweden recorded from Nb., Lu.Lpm. and T.Lpm.; in Norway recorded from Fn, Fø, On, STi; in Finland found in ObS, Ks, LkW, Le and Li; also on the Kola Peninsula. – Otherwise there are no records of the species.

Biology. Not much information about the biology is available; Angus (1970d) mentions the species from gravelly ditches with *Sparganium* and other plants, or mosses (not *Sphagnum*), in damp moorland, and from a flooded field. According to Muona (*in litt.*) it occurs (in NE. Finland) in open, wet *Sphagnum*-pools. Adults are found in June-August.

Subgenus *Atracthelophorus* Kuwert, 1886

Atractohelophorus Kuwert, 1886b, Wien. ent. Ztg. 5: 227.
Atracthelophorus Kuwert, 1890, Bestimm. Tab. eur. Col. 20: 25 (emend.).
   Type species: *Helophorus arvernicus* Mulsant, 1846, by subsequent designation (Sharp, 1915a).

Elytra without intercalary stria. 11th interstice strongly and sharply ridged; pseudepipleura visible from below, normally as wide as epipleura (opposite metacoxae). Terminal segment of maxillary palpi symmetrical oval (Figs 145, 146); in some specimens of *glacialis,* however with a little straighter inner margin (see also note under the following subgenus). Tarsi with fine swimming hairs on their dorsal face. Antennae 9-segmented.

38. *Helophorus brevipalpis* Bedel, 1881
   Figs 145, 151, 184.

Helophorus brevipalpis Bedel, 1881, Faune Col. bass. Seine 1: 301.
Helophorus guttulus auctt.; *nec* Motschulsky, 1860.

2.4-3.2 mm. A variable species. Piceous to black, anterior and lateral margins of pronotum mostly yellowish. Metallic reflections of head and pronotum purplish, bronze or greenish, often bronze with middle and (especially) internal intervals of pronotum bright green. Elytra yellow or brownish yellow, with distinct black arrow-mark, a black spot opposite this on 7th interstice, and some vaguely defined paler spots, often somewhat mottled. Granulation of head and pronotum variable, mostly rather feeble and reduced, but the granules usually detectable. Frontal furrow expanded anteriorly. Pronotum moderately convex, lateral margins straight or only slightly sinuate in posterior half (Fig. 151). Elytra strongly striate, interstices uniform, feebly convex. Appendages reddish yellow; maxillary palpi often brownish with darker apex; antennal club darker; claw segments darker distally. Shape of terminal segment of maxillary palpi variable, short or long, but always symmetrical.
   ♂: Aedeagus, Fig. 184.

   Distribution. A very common and widespread species; the whole of Denmark and Sweden, north to Ly.Lpm. and Nb.; in Norway north to NSy (only scattered records from the mountainous regions in the south); whole Finland north to LkW and LkE. – Widely distributed; North Africa, entire Europe (with the exception of northernmost Fennoscandia), Asia Minor and the Caucasus, eastwards to the Urals; also introduced to North America (USA).

Biology. Usually in stagnant fresh water, preferring shallow, more or less open, often temporary pools with grassy bottom, yet fairly euryoecious, found under rather different conditions; also in slower reaches of running waters. Life cycle as in *aequalis*. Mainly in spring and autumn, often very abundant. The species is a very active flyer, frequently found in drift on the seashore.

39. *Helophorus arvernicus* Mulsant, 1846
Figs 146, 152, 185

Helophorus arvernicus Mulsant, 1846, Hist. Nat. Col. Fr. Suppl.: 281.

2.7-3.5 mm. A rather short and convex species. Piceous to black, anterior margin of pronotum narrowly paler. Head and pronotum metallic, reddish or purplish bronze, the granules mostly greenish. Elytra brown or yellowish brown, with some indistinct paler spots, and with distinct black arrow-mark, and a black spot opposite this on 7th interstice. Head and pronotum closely and coarsely granulate, the granules distinct. Frontal furrow expanded anteriorly. Pronotum strongly convex, lateral margins strongly rounded anteriorly, rather strongly sinuate posteriorly (Fig. 152). Elytra rather convex, very strongly striate; 1st, 3rd, 5th and (usually) 7th interstices raised above the others. Maxillary palpi brown or yellowish brown; antennae and legs yellowish brown, the former with darker club.
♂: Aedeagus, Fig. 185.

Distribution. In Denmark known only from Funen and Jutland, where it is widespread and not uncommon; in Finland only recorded from Sa; not in Norway and Sweden. – Ranges from West Europe to the European USSR, from southern Fennoscandia, the Kola Peninsula and the Baltic region of USSR to North Spain, North Italy and Hungary.

Biology. In running fresh water, particularly at the grassy edges of smaller, softbottomed streams; in shallow water among vegetation or in wet mud on the banks. The species show a very poor ability of dispersal and is perhaps unable to fly; contrary to most other *Helophorus*-spp. it is not found in sea drift. Egg cocoons are placed in mud at the edge of the water. Adults are found mainly in spring, but also in summer and autumn.

40. *Helophorus glacialis* Villa, 1833
Fig. 183.

Elophorus glacialis Villa, 1833, Col. Eur. Dupl. (1): 34.

2.5-4.0 mm. Black; head and pronotum metallic, blackish green or blackish purple; pronotum sometimes purplish bronze laterally and in the furrows. Elytra with faint, but mostly distinct, greenish, bronze or purplish hue, usually very dark, blackish or brown, sometimes with indistinct paler spots, or even pale brown with dark spots.

Head and pronotum with simple punctuation, or at most with strongly reduced granulation, only external intervals of pronotum distinctly granulate. Frontal furrow rather narrow, only slightly expanded anteriorly. Pronotum relatively small, lateral margins straight or slightly sinuate in posterior half. Elytra with an oblique depression in anterior third, finely striate, interstices uniform, hardly convex. Appendages very dark, usually black; legs and maxillary palpi with distinct green or purplish bronze reflections, the latter strongly metallic. Terminal segment of maxillary palpi sometimes slightly asymmetrical, with straighter inner face (as in the following subgenus), but the species is easily recognized by the metallic legs and palpi.

♂: Aedeagus, Fig. 183.

Distribution. A boreomontane species; widespread in North and Central Sweden south to Dlr.; in Norway very widespread and common; in Finland only in the north (Li, Le); not in Denmark. – Europe; besides Fennoscandia the species occurs in the mountains further south, ranging from Spain, through the Alps and the Appennines to the Carpathians and the Balkans.

Biology. In stagnant water, very stenothermic, apparently confined to very cold water at the edge of snow patches (where the water is always near freezing-point); usually in small and shallow dark-bottomed pools left by melting snow, apparently preferring pools with stony or clayey bottom and rich in mosses. It is found in spring and July-August, often very abundant.

Subgenus *Helophorus* Fabricius, 1775, s.str.

Elophorus Fabricius, 1775, Syst. Ent.: 66.
Helophorus Illiger, 1801 (emend.).
  Type species: *Elophorus minutus* Fabricius, 1775 (see above under the generic description).

Differs from the preceeding subgenus only by the shape of the terminal segment of the maxillary palpi, which is distinctly asymmetrical, with straighter inner face (Fig. 149). The pseudepipleura may be either visible from below (about as wide as the epipleura) or narrow and not visible from below. Antennae 8- or 9-segmented.
  The distinction between this subgenus and *Atracthelophorus* is not well-founded from a strict phylogenetic view, and merely ought to be considered as practical.

41. *Helophorus granularis* (Linnaeus, 1761)
    Figs 158, 175, 187.

Buprestis granularis Linnaeus, 1761, Fauna Suec. (2): 214.

2.2-2.8 mm. Black or pitchy brown, margins of pronotum hardly lighter. Head and pronotum with dark, greenish or purplish bronze metallic reflections. Elytra yellowish brown, at middle with a black arrow-mark and a black spot opposite this on 7th interstice, without lighter mottling. Head with feeble, but mostly distinct granulation; frontal furrow expanded anteriorly. Pronotum (Fig. 158) with close and usually coarse granulation, the granules often very low, but normally distinct also on internal intervals. Pronotum transversely rather strongly and evenly convex (almost to lateral margins), hardly convex longitudinally; widest just before middle, lateral margins rather evenly curved; submedian furrows feebly curved outwards in middle, but seldom angular; marginal furrow narrow. Elytra strongly striate, interstices fairly strongly convex. Appendages reddish yellow, antennal club brownish, apex of maxillary palpi and distal part of tarsi darker (blackish).

σ: Aedeagus, Fig. 187.

The species may resemble *brevipalpis* but differs by the asymmetrical terminal segment of the maxillary palpi and the narrow pseudepipleura, that are not visible from below. Recognized from the similar *redtenbacheri* by the 9-segmented antennae, finer elytral striae, the narrow pseudepipleura, and the anteriorly expanded frontal furrow. For separation from the also very similar *discrepans* and *minutus,* see under those species.

Distribution. Widely distributed in Fennoscandia and Denmark; found in the whole of Denmark (rare!) and Sweden; widespread in southern Norway north to STi; in Finland north to ObS and Ks. – A very widespread species, known from entire Europe (except northern Fennoscandia); eastwards to Siberia.

Biology. In stagnant, sometimes also slowly flowing waters, perhaps preferring acid water and mainly pools with clayey or sandy bottom and abundant vegetation at the edges; also in temporary pools. Life cycle as in *aequalis.* Mainly found in spring and autumn, often (particularly in spring) taken in drift on the seashore. The species is usually macropterous, but a brachypterous form (var. *ytenensis* Sharp) is described from the British Isles.

42. *Helophorus discrepans* Rey, 1885
Fig. 188.

Helophorus discrepans Rey, 1885, Annls Soc. linn. Lyon 31 (1884): 19.

2.8-3.3 mm. Very similar to both *granularis* and *minutus,* in external characters being almost intermediate between the two. Body shape a little more elongate than in *granularis;* submedian furrow of pronotum bluntly angled outwards in middle (in *granularis* seldom angular). From *minutus* it differs in having pronotum slightly more arched (though less strong than in *granularis*), more strongly granulate, and without broad pale margins. Terminal segment of maxillary palpi normally slightly shorter than in *minutus.* However, the external characters show some variation (in all three species) so examination of the aedeagus which is quite distinct (Fig. 188) can be necessary.

Distribution. Not in Denmark, Sweden and Norway, and not with certainty found in Finland. It is known from Vib. in the USSR, and according to Angus (1974) it is possible that the specimens recorded as *walkeri* from Finland (Kangas, 1968b) belong to *discrepans* (see also note under *H. obscurus*). – East Europe, from South-east Germany, West Poland and Estonia eastwards to the Caucasus, from Leningrad to Greece; also in the Pyrenees.

Biology. In stagnant water, perhaps mainly in shallow, temporary pools, e.g. found in spring in pools left by melting snow. In the more western parts of its area of distribution it inhabits mountain streams and upland pools.

43. *Helophorus minutus* Fabricius, 1775
Figs 149, 159, 176, 186, 189; pl. 2: 5.

Elophorus minutus Fabricius, 1775, Syst. Ent.: 66.

2.5-3.6 mm. Black or pitchy brown, anterior and lateral margins of pronotum pale yellow. Head and pronotum metallic, usually bright green or bronze. Elytra yellow, mostly rather pale, in middle with a blackish arrow-mark, and a dark spot opposite this on 7th interstice. Granulation of head and pronotum finer than in *granularis,* especially on clypeus and internal intervals of pronotum, feeble and often reduced (but the granules mostly detectable). Frontal furrow expanded anteriorly. Pronotum (Fig. 159) moderately convex transversely, not convex longitudinally, widest about anterior third, lateral margins usually a little straightened posteriorly, but not sinuate. Elytral striae a little finer than in *granularis,* interstices only moderately convex. Pseudepipleura narrow, not visible from below. Appendages reddish yellow; antennal club, apex of maxillary palpi, and distal part of claw segments darker. Metatarsal claw segment (exclusive of claws) not longer than 1st + 2nd segment. Antennae 9-segmented.
 ♂: Aedeagus, Fig. 189.
 For separation from *discrepans, lapponicus, griseus, longitarsis* and *fulgidicollis,* see under those species.

Distribution. Fairly common in southern Scandinavia; entire Denmark; most Swedish provinces north to Dlr.; in Norway only mentioned from the south (Ø, AK, Os, Bø, VE); southern Finland (Al, Ab, N, Ka, Ta). – Widely distributed; North Africa, entire Europe north to Central Fennoscandia, eastwards to West Siberia.

Biology. In stagnant, only occasionally running, fresh water, mainly in shallow, open, eutrophic pools with grassy and usually clayey bottom, particularly in rather light-bottomed pools. Life cycle as in *aequalis.* Mainly in spring and autumn. The species is an active flyer, frequently found in drift on the seashore.

Note. Lindroth (1960) gives records of *minutus* from northern Sweden (Med., Nb.), but according to A. Nilsson (*in litt.*) its occurrence there is most doubtful. Specimens from these provinces standing in museum collections appear to be misidentified specimens of other species (*granularis, flavipes*).

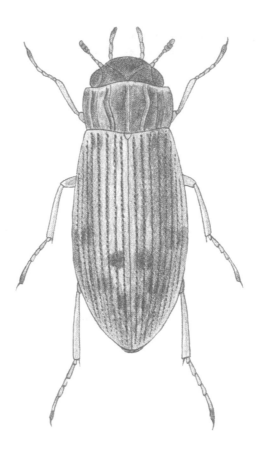

Fig. 186. *Helophorus minutus* F., length 2.5-3.6 mm. (After Victor Hansen).

44. *Helophorus lapponicus* Thomson, 1853
Fig. 192.

Helophorus lapponicus Thomson, 1853, Öfvers. K. VetenskAkad. Förh. 10: 42.
Helophorus celatus Sharp, 1916, Entomologist's mon. Mag. 52: 171.
Helophorus erichsoni auctt.; *nec* Bach, 1866.

3.3-4.0 mm. A rather variable species; very similar to *minutus,* of the same colour, though yellow margins of pronotum may be very dull. Frontal furrow expanded anteriorly. Pronotum varying from transversely highly arched to almost flat (hardly convex longitudinally); the granulation may be coarse over all intervals, or weak, or even

112

absent from internal intervals; the lateral margins curved as in *minutus,* or in posterior half straighter, but only very seldom with a suggestion of sinuation. Elytra on the average slightly longer than in *minutus,* fairly strongly striate (almost as in *minutus*). Pseudepipleura narrow, hardly visible from below. Metatarsal claw segment (exclusive of claws) distinctly longer than 1st + 2nd segment. Antennae 9-segmented.

σ: Aedeagus, Fig. 192.

The species is separated from similar species (except *fulgidicollis*) by the long metatarsal claw segment. Furthermore it differs from *granularis* and *discrepans* by the larger size, from *griseus* and *longitarsis* by the stronger elytral striae, from *griseus* also by the shape of pronotum, and from *longitarsis* by the (at least laterally) distinct granulation on pronotum; for separation from *fulgidicollis,* see under that species.

Distribution. A boreomontane species; widespread in North and Central Fennoscandia; in Sweden south to Dlr. (also an isolated occurrence on Öl.); in Norway known from TRi, Fi, Fn, Fø in the north, and HEn, On, STi further south; in Finland found in most provinces south to Tb and Sb (also found in N); from Denmark only two specimens, found in drift on the seashore at Bøtø (LFM). – An eastern species, ranging from Fennoscandia through USSR (where it is widespread) to Siberia and Mongolia; isolated populations occur in North Spain (Cantabrian mountains) and in the Caucasus.

Biology. In stagnant fresh water, apparently preferring shallow, sparsely vegetated, and somewhat acid, temporary pools, surrounded by mosses (*Polytrichium*), often together with e.g. *H. laticollis.* Life cycle almost as in *aequalis;* egg cocoons (containing about 6 eggs) are however placed in shallow water among submerged vegetation. Mainly found in spring and early summer; also taken in drift on the seashore, cf. above.

45. *Helophorus fulgidicollis* Motschulsky, 1860
Figs 155, 177, 193.

Helophorus fulgidicollis Motschulsky, 1860, Reis. Forsch. Amur-L. Schrenk 2 (2): 105.

3.2-4.5 mm. Piceous to black, anterior and lateral margins of pronotum yellowish. Metallic reflections of head and pronotum usually darker than in *minutus,* head mostly greenish or bronze; pronotum normally greenish, laterally and in the furrows orange or purplish bronze. Elytra yellowish brown, darker than in the preceeding species of the subgenus, the arrow-mark often indistinct. Head and pronotum evenly, strongly and closely granulate, though the granules are usually low. Frontal furrow expanded anteriorly. Pronotum (Fig. 155) widest in about anterior third, lateral margins usually relatively weakly rounded anteriorly, and straight, or almost so posteriorly (not sinuate). Elytra strongly striate, interstices strongly convex, the alternate interstices often raised a little above the others. Pseudepipleura narrow, hardly visible from below. Appendages yellowish red; apex of maxillary palpi and apex of claw segments darker (brownish). Metatarsal claw segment (exclusive of claws) distinctly longer than 1st + 2nd segment (Fig. 177). Antennae 9-segmented.

♂: Aedeagus, Fig. 193.

Distinguished from the similar species (except *lapponicus*) by the long metatarsal claw segment; from *lapponicus* (and the other similar species) it differs by the relatively dark and strongly striate elytra. Swimming hairs of meso- and metatarsi longer than in the similar (pale) species.

Distribution. In Denmark not uncommon along the coasts (not B); in Sweden only known from a few localities in south-western Sk.; not in Norway and Finland. – Eu-

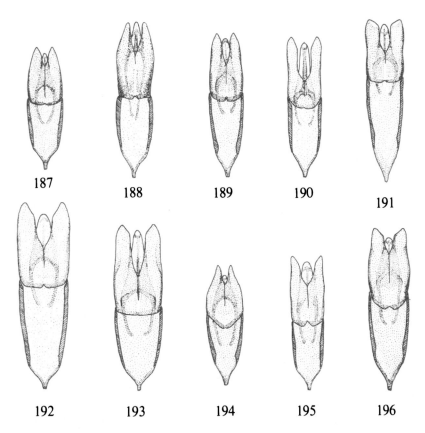

Figs 187-196. Male genitalia of *Helophorus* (dorsal view). – 187: *granularis* (L.); 188: *discrepans* Rey; 189: *minutus* F.; 190: *griseus* Hbst.; 191: *longitarsis* Woll.; 192: *lapponicus* Thoms.; 193: *fulgidicollis* Motsch.; 194: *nanus* Sturm; 195: *redtenbacheri* Kuw.; 196: *pumilio* Er.

114

rope, exclusively along the coasts; from South Scandinavia to the English Channel and the Irish Sea, the Atlantic coast (Lisbon) and the western Mediterranean (including Morocco), eastwards to West Italy.

Biology. A halobiontic species, predominantly in small, shallow, often temporary and sparsely vegetated, brackish pools on the salt marshes above high tide line. The species is a comparatively good swimmer (though it should not be compared to *Berosus*-spp. or swimmmers like these), and can usually be seen swimming freely over the bottom. But it is obviously not an active flyer; it has not been found in sea drift. Life cycle as in *aequalis*. Adults are found mainly in spring and autumn, often very numerously.

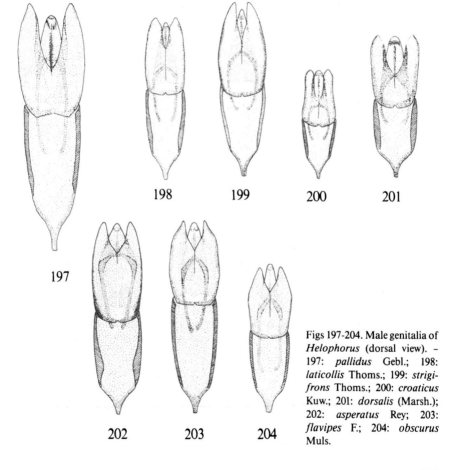

198   199   200   201

197

Figs 197-204. Male genitalia of *Helophorus* (dorsal view). – 197: *pallidus* Gebl.; 198: *laticollis* Thoms.; 199: *strigifrons* Thoms.; 200: *croaticus* Kuw.; 201: *dorsalis* (Marsh.); 202: *asperatus* Rey; 203: *flavipes* F.; 204: *obscurus* Muls.

202   203   204

115

46. *Helophorus griseus* Herbst, 1793
Figs 160, 190.

Elophorus griseus Herbst, 1793, Natursyst. all. bek. in- u. ausl. Ins. 5: 143.
Helophorus semifulgens auctt.; *nec* Rey, 1885.

2.8-3.8 mm. Very similar to *minutus* in colour and size, but the metallic reflections of head and pronotum are often brighter; these may very seldom be indistinct, the pronotum being almost black, but still with distinct yellow margins. Granulation of head and pronotum more reduced than in *minutus,* internal intervals of pronotum often with simple punctuation; lateral margins of pronotum more straight in posterior half and before posterior angles distinctly (though often very feebly) sinuate (Fig. 160). Elytra on the average a little wider than in *minutus* and (especially) *lapponicus,* the striae a little finer. Frontal furrow expanded anteriorly. Pseudepipleura narrow, not visible from below. Metatarsi as in *minutus* (only extremely rarely with longer claw segment, as in *lapponicus*).

♂: Aedeagus, Fig. 190.

For separation from the very similar *longitarsis,* see under that species.

Distribution. Southern Scandinavia; fairly common in Denmark; scattered records from southern Sweden (Sk., Öl., Gtl., Vg., Vrm.); in Norway only recorded from Ø, AK and Bø; in Finland only known from Al. – Throughout Europe, north to southern Fennoscandia; eastwards to the Caucasus (perhaps further east).

Biology. As *minutus,* often found together with it.

Helophorus longitarsis Wollaston, 1864
Fig. 191.

Helophorus longitarsis Wollaston, 1864, Cat. Col. Ins. Canar.: 86.
Helophorus erichsoni Bach, 1866, Syst. Verz. Käf. Deutschl.: 11.
Helophorus semifulgens Rey, 1885, Annls Soc. linn. Lyon 31 (1884): 388.

3.0-4.0 mm. Very similar to *minutus* and *griseus,* particularly the latter. Anterior margin and marginal furrows of pronotum broadly yellow; elytra pale yellow. Pronotum shining, metallic reflections bright, greenish, bronze or purplish, its granulation more reduced than in any of the preceeding species of the subgenus; middle and internal intervals with fine and simple punctuation, external intervals stronger punctured, and often somewhat corrugate, but without distinct granules; lateral margins of pronotum straight or weakly rounded in posterior half, not sinuate. Elytra on the average slightly wider than in *griseus,* the striae as in *griseus.*

♂: Aedeagus, Fig. 191.

Distribution. So far not recorded from Scandinavia. – Europe; ranging from Great Britain and Spain to the Urals, from Italy and Greece to North Germany (Holstein); in Russia only in the south. Possibly it may also occur in South Scandinavia.

Biology. As *minutus,* yet apparently more thermophilic.

47. *Helophorus nanus* Sturm, 1836
Figs 157, 163, 174, 194.

Elophorus nanus Sturm, 1836, Deutschl. Ins. 10: 40.

2.3-2.8 mm. Black; dorsal face very shining; head and pronotum dark metallic, greenish, purplish or bronze; elytra dark uniformly brown or paler brown with a variable, usually large, dark sutural spot in middle, and a smaller spot opposite this on 7th interstice. Pronotum (Fig. 157) highly and evenly arched, convex both transversely and longitudinally, widest about middle, the lateral margins evenly curved. Head and pronotum very shining, smooth, often almost polished, with very fine punctuation, often without any trace of granulation, except towards the lateral margins. Frontal furrow narrowly linear; between it and the eyes on each side with a distinct (though sometimes rudimentary) accessory short longitudinal furrow (Fig. 163). Elytra strongly striate, interstices rather convex. Pseudepipleura broadly visible from below. Maxillary palpi and antennae yellowish red, apex of the former darker; legs reddish brown with distal part of claw segments darker. Antennae 8-segmented (Fig. 174).
σ: Aedeagus, Fig. 194.

Distribution. Fairly common in Denmark; most Swedish provinces north to Vstm. (also recorded from Nb. and T. Lpm.); widespread in Finland (though not recorded from some provinces in Central Finland); in Norway only mentioned from Bø and Fi. – Widely distributed throughout North and Central Europe, ranging from Ireland and France through Siberia and Manchuria to the Pacific coast (Vladivostok), from Fennoscandia to North Italy.

Biology. In stagnant fresh water, especially eutrophic, shallow, often temporary pools with clayey and grassy bottom, both in open country and in woodland. Life cycle as *aequalis*. Mainly in spring, often very abundant; sometimes also in drift on the seashore.

48. *Helophorus redtenbacheri* Kuwert, 1885
Fig. 195.

Helophorus pumilio Thoms. (sic) var. *redtenbacheri* Kuwert, 1885, Wien. ent. Ztg. 4: 264.
Helophorus pumilio auctt.; *nec* Erichson, 1837.

2.4-3.0 mm. Similar to *nanus* in colour and size; elytra normally dark, but sometimes paler with dark markings; metallic hue of head and pronotum often very dark. Head feebly, though mostly distinctly granulate; frontal furrow narrowly linear, only very seldom slightly expanded anteriorly. Pronotum rather closely and coarsely, almost evenly granulate, the granules low, but fairly large. Shape of pronotum almost as in *nanus*, highly arched, strongly convex transversely, distinctly convex longitudinally (but normally not as strong as in *nanus*). Elytra strongly striate, interstices strongly convex (almost as in *pumilio*). Pseudepipleura broadly visible from below. Antennae

8-segmented. Paler specimens with frontal furrow expanded anteriorly may resemble *granularis,* but are easily recognized by the 8-segmented antennae, the wide pseudepipleura and the stronger elytral striae.

♂: Aedeagus, Fig. 195.

Distribution. Very rare; with certainty only known from a few localities in southern Scandinavia; in Denmark SZ (Knudshoved) and NEZ (Damhusmosen, Amager-fælled); in Sweden only Sk.; not in Norway and Finland. – Eastern Europe, from West Germany to West Siberia, from southern Scandinavia to the northern Balkans.

Biology. In stagnant water, in shallow, often temporary, eutrophic pools with grassy and clayey bottom, on open ground near the coast. The restriction (in Scandinavia) to coastal areas is probably correlated with climatic factors, rather than a certain salinity of the inhabited waters; it is not at all halobiontic, nor halophilic, and does not occur on the salt marshes. Mainly found in March-April.

Note. Many of the Swedish specimens standing as *redtenbacheri* (earlier referred to as *pumilio*) appear to be misidentified specimens of other species, mainly *granularis.* Besides a few specimens from Sk., it was not possible to confirm any of the Swedish records (Hall., Sdm. and Upl., as mentioned by Lindroth, 1960), which are therefore omitted.

49. *Helophorus pumilio* Erichson, 1837
Fig. 196.

Helophorus pumilio Erichson, 1837, Käf. Mark Brand. 1: 197.
Helophorus fallax Kuwert, 1886a, Wien. ent. Ztg. 5: 135.

2.5-3.2 mm. Similar to *nanus* in colour and size, normally with uniformly dark elytra. Granulation of head and pronotum feeble, on internal intervals of pronotum usually reduced to simple punctuation, middle portion of pronotum therefore shining, often almost polished (as in *nanus*). Frontal furrow narrow linear, between it and the eyes often with accessory furrows (almost as in *nanus,* but weaker). Pronotum widest in, or just behind middle, the sides evenly, rather strongly curved, more strongly and highly arched than in *nanus* and *redtenbacheri,* markedly convex longitudinally, internal intervals just behind middle slightly more convex (somewhat humpy). Elytra strongly striate, interstices strongly convex. Pseudepipleura broadly visible from below. Antennae 9-segmented.

♂: Aedeagus, Fig. 196.

Distribution. Only found in Denmark: Østrig near Tranekær (F), several specimens on the seashore, 3.6. and 10.6. 1979 (under seaweed); not in the rest of Fennoscandia. – Europe; from Central France to Siberia, from the Alps and the northern Balkans to Denmark and the Yaroslavl area (USSR). (According to Angus (*in litt.*) it is common on the southern side of the Gulf of Finland).

Biology. In the west (e.g. France and southern Germany) the species is normally found in streams, but farther east it mainly inhabits small pools, e.g. temporary pools left by melting snow. It is also found in drift on the seashore, cf. above. Perhaps mainly in May-June.

50. *Helophorus pallidus* Gebler, 1830
Fig. 197; pl. 2: 8.

Helophorus pallidus Gebler, 1830, Ledebours Reise, Ins. 2: 103.

4.5-5.0 mm. A characteristic, rather oblong and narrow species. Paler than the other species with narrow frontal furrow. Reddish brown or brown; head and pronotum with metallic, purplish or bronze, sometimes indistinct reflections, elytra yellowish or reddish brown, usually with a darker, often indistinct sutural spot in middle. Head and pronotum closely and evenly granulate, the granules distinct, but very low, only on clypeus indistinct. Frontal furrow narrowly linear, between it and the eyes with short accessory furrows (as in *nanus*). Pronotum moderately strongly and fairly evenly arched, only feebly convex longitudinally (or hardly so), widest in about anterior third, the sides straight posteriorly. Elytra rather narrow, 2/3 longer than wide, fairly strongly striate, interstices convex (especially the alternate interstices). Pseudepipleura either broadly visible from below, or narrower, sometimes hardly apparent. Appendages brownish or reddish yellow, apex of maxillary palpi and apex of claw segments darker. Antennae 8-segmented.

♂: Aedeagus, Fig. 197.

Distribution. Widespread but rather uncommon in northern Sweden, south to Jmt. and Äng.; North Finland south to Kb.; in Norway only recorded from TRi; not in Denmark. – Northern Palearctis, ranging from NE. Fennoscandia and Estonia to East Siberia.

Biology. Apparently to some extent a tundra and steppe species (Angus, 1974); probably preferring small, shallow pools. June, August.

51. *Helophorus laticollis* Thomson, 1853
Fig. 198.

Helophorus laticollis Thomson, 1853, Öfvers. K. VetenskAkad. Förh. 10: 43.

3.0-4.0 mm. Similar to *strigifrons* in size and colour; legs and maxillary palpi however usually slightly darker. Granulation of head and pronotum close and distinct, but mostly rather feeble; the granules low, especially on clypeus and internal intervals of pronotum. Frontal furrow narrowly linear. Pronotum highly and evenly arched, distinctly convex longitunally, internal intervals not humpy (as they are in *strigifrons*); lateral margins more evenly curved than in *strigifrons,* usually not straight in posterior third; its furrows more superficial than in *strigifrons.* Pseudepipleura not visible from below. Antennae 9-segmented.

♂: Aedeagus, Fig. 198.

Distribution. In Denmark very rare, only a few localities from SJ, WJ and NEJ; widespread, but sporadic throughout Sweden; widespread in southern Norway; in Finland very widespread, north to LkW and LkE. – North-east European species, ranging from SE. England to the Yaroslavl area (USSR), from North Finland and the Kola Peninsula to Central Germany.

Biology. In stagnant fresh water, mainly in shallow, acid and normally oligotrophic, temporary pools with sparse vegetation. Life cycle almost as in *aequalis;* egg cocoons are however placed in shallow water among submersed vegetation (as in *lapponicus*). Perhaps mainly in spring and summer, but also in autumn.

52. *Helophorus strigifrons* Thomson, 1868
Figs 156, 164, 199.

Helophorus strigifrons Thomson, 1868, Skand. Col. 10: 308.

3.5-4.2 mm. Pitchy brown or black, lateral margins of pronotum narrowly paler, head and pronotum metallic, blackish green, dark bronze or dark purplish. Elytra brown with darker, indistinct sutural arrow-mark. Head and pronotum closely and rather coarsely granulate, granulation on clypeus and middle portion of pronotum less prominent and often strongly reduced, internal intervals of pronotum normally shining. Frontal furrow narrowly linear, rather deep (Fig. 164). Pronotum (Fig. 156) widest in about anterior third or just before middle, the sides smoothly curved, often straightened in posterior third; strongly, somewhat unevenly arched, only feebly (or hardly) convex longitudinally, but with internal intervals strongly convex in about middle, more or less humpy. Elytra fairly strongly striate, interstices rather convex. Pseudepipleura broadly visible from below. Appendages reddish brown to yellowish red; apex of maxillary palpi, antennal club and distal part of claw segments darker. Antennae 9-segmented.
♂: Aedeagus, Fig. 199.

Distribution. Common and widespread throughout most of Denmark and Fennoscandia; entire Denmark; entire Sweden; in Norway widely distributed (though not mentioned from a number of south-western provinces), north to Fi; in Finland very widespread and found in almost all provinces. – Widely distributed throughout North and Central Europe, ranging from Great Britain and NE. France to Siberia, from Fennoscandia to South Germany.

Biology. In stagnant, mostly eutrophic fresh water, normally in small, grassy pools, often in woodland, yet avoiding the more shaded sites. Life cycle as in *aequalis.* March-October, mainly in the spring.

Helophorus croaticus Kuwert, 1886
Figs 165, 200.

Helophorus strigifrons var. *croaticus* Kuwert, 1886b, Wien. ent. Ztg. 5: 248.

120

2.7-3.8 mm. A rather short and convex species, in general appearance somewhat similar to *arvernicus* (subgenus *Atracthelophorus*), but with differently shaped pronotum. Black or pitchy brown, lateral margins of pronotum narrowly reddish. Head and pronotum with metallic, purplish bronze and (on the granules) greenish reflections; elytra yellowish brown to brown, with distinct dark sutural arrow-mark, and distinct lighter mottling; at least with a distinct pale spot in 3rd interstice (just before arrow-mark) and one in 5th interstice (in about anterior quarter). Granulation of head and pronotum distinct, but feeble. Frontal furrow subparallel (Fig. 165), hardly expanded anteriorly, about twice as wide as in *strigifrons*. Pronotum moderately arched, hardly convex longitudinally, internal intervals slightly more convex (but hardly humpy); widest in about anterior third, the sides almost straight in posterior third, not sinuate. Elytra rather short, only about 1/2 longer than wide, very strongly striate, interstices strongly convex, in part obtusely ridged. Pseudepipleura broadly visible from below, opposite metacoxae about as wide as epipleura. Appendages yellowish red; apex of maxillary palpi, antennal club, and distal part of claw segments darker. Antennae 9-segmented.

♂: Aedeagus, Fig. 200.

Distribution. So far not recorded from Scandinavia. – An eastern species; Central Europe, widely distributed in the European USSR; in Central Europe northwards to Belgium and NW. Germany (Lohse *in litt.),* towards east extending farther north (to Yaroslavl). Also found in East Siberia (Yakutsk); and probably widespread over North Siberia. Perhaps the species may be found in southernmost Scandinavia or in East Fennoscandia.

Biology. Not much information about the biology is available. Probably it prefers stagnant fresh water; in East Siberia the species is found in weedy and grassy pools beside the river Lena, and in Bavaria in backwaters of a river (Angus, *in litt.).* Adults are found in July.

Helophorus dorsalis (Marsham, 1802)
Fig. 201.

Hydrophilus dorsalis Marsham, 1802, Ent. Brit. 1: 410.

3.0-3.8 mm. A characteristic species, recognized from the other species in the subgenus by the extensive lighter mottling of the elytra. Black to pitchy brown, lateral margins of pronotum yellowish red; metallic reflections of head and pronotum purplish bronze, outside furrows often greenish. Elytra yellowish brown, with distinct dark sutural arrow-mark, and with pronounced lighter mottling; each elytron with a large, very conspicuous pale spot behind arrow-mark (covering interstices 2-5), a small pale spot just anterior to arrow-mark (covering at least interstice 3), and a rather large spot in anterior third (covering interstices 4-5); usually also with other, less conspicuous paler spots. Head and pronotum feebly granulate, clypeus and internal intervals of pronotum shining, with only reduced and partially indistinct granulation. Frontal furrow expanded anteriorly, wider than in *croaticus*. Pronotum widest in about anterior third, narrowed posteriorly in almost straight lines, rather evenly arched, not distinctly convex longitudinally; internal intervals slightly stronger convex in middle, but not humpy. Elytra longer than in *croaticus,* about 3/5 longer than wide, the striae rather strong, interstices rather convex (but much weaker than in *croaticus*). Pseudepipleura broadly visible from below. Appendages yellow or reddish yellow; apex of maxillary palpi, antennal club, and distal part of claw segments darker. Antennae 9-segmented.

σ: Aedeagus, Fig. 201.

Distribution. So far not recorded from Scandinavia. – South and Central European species, perhaps mainly in the western region; ranging from France and the British Isles eastwards to western Ukraine, northwards to the Netherlands and NW. Germany (Holstein). Possibly the species may also be found in South Scandinavia.

Biology. In stagnant fresh water, apparently preferring shallow grassy pools or ponds with clayey bottom on somewhat shaded sites in woodland. July.

53. *Helophorus asperatus* Rey, 1885
Figs 161, 202.

Helophorus asperatus Rey, 1885, Annls Soc. linn. Lyon 31 (1884): 19.

3.8-4.8 mm. Black, anterior and lateral margins of pronotum reddish. Head and pronotum metallic, greenish or purplish bronze. Elytra brown or yellowish brown, with a very weak (often almost indistinct) normally greenish metallic hue, at least on 1st interstice; in middle with a darker sutural arrow-mark, and a dark spot opposite this on 7th interstice. Head and pronotum evenly, strongly and closely granulate, the granules very distinct and strikingly well defined, also on internal intervals of pronotum, showing no trace of fusing. Frontal furrow usually rather narrow, linear or sublinear (Fig. 161). Pronotum widest in about anterior third, the sides straight or only weakly rounded in posterior half. Elytra with a weak, sometimes almost disappearing, oblique depression in anterior third; the striae rather strong, interstices rather convex. Pseudepipleura narrow, hardly visible from below. Appendages reddish to yellowish brown, apex of maxillary palpi and apex of claw segments darker. Antennae 9-segmented.

σ: Aedeagus, Fig. 202.

The species may resemble *fulgidicollis,* but differs by the wider, posteriorly more strongly narrowed pronotum with stronger and more well defined granules, the faint metallic hue on the elytra, and (usually) narrower frontal furrow; these characters (except for the narrow frontal furrow) also distinguish the species from *strigifrons;* from *flavipes* and *obscurus* it differs by the weaker metallic hue and weaker depression on elytra, slightly longer swimming hairs of meso- and metatarsi, and normally by the strong pronotal granulation (some very rare specimens of *obscurus* may be strongly and evenly granulate, as in *asperatus;* such specimens are easily distinguished by the aedeagus).

Distribution. Very rare, only found singly at a few localities in southernmost Scandinavia; in Denmark only SJ (Draved Skov, Sønderborg), WJ (Esbjerg), NWJ (Vust) and B (Rønne); from Sweden only one specimen from Sk. (Vitemölla); not in Norway and Finland. – Central Europe, from France to Poland, from South Scandinavia to North Italy.

Biology. In stagnant, and perhaps slowly flowing fresh water. Not much information about the habitat of the species is available; most of the (very few) Scandinavian specimens were found in drift on the seashore. May-June, September.

54. *Helophorus flavipes* Fabricius, 1792
Figs 142, 153, 162, 203; pl. 2: 6.

Elophorus flavipes Fabricius, 1792, Ent. Syst. 1 (1): 205.
Helophorus viridicollis Stephens, 1829, Ill. Brit. Ent. Mandib. 2: 112.

3.0-4.0 mm. A variable species, normally rather easily recognized from all the preceeding species of the subgenus (except *asperatus*) by the metallic hue on the elytra. Black, lateral margins of pronotum sometimes dull reddish. Head and pronotum dark metallic, normally blackish or brownish green or dark bronze, in the furrows usually more purplish. Elytra brown or dark brown, with a distinct, though sometimes very faint metallic hue of the same colour as that of head and pronotum; the sutural arrow-mark only distinct in paler specimens. Head and pronotum closely granulate, the granulation on clypeus and internal intervals (often also middle intervals) of pronotum reduced and very feeble, often reduced to simple punctuation. Frontal furrow expanded anteriorly (Fig. 162); submedian furrow of pronotum curved outwards in middle, but often only weakly so, not sharply angled (Fig. 153). Elytra with a distinct shallow, oblique depression in anterior third; the striae moderately strong, interstices usually only moderately convex. Elytra normally tapered at apex. Pseudepipleura narrow, normally not visible from below (Fig. 142). Appendages reddish or yellowish brown, apex of maxillary palpi, antennal club and distal part of claw segments darker (blackish or brown). Swimming hairs of meso- and metatarsi rather short. Antennae 9-segmented.
♂: Aedeagus, Fig. 203.

Distribution. A common and very widespread species, found throughout Denmark and Fennoscandia. – Widely distributed; entire Europe, eastwards to Siberia.

Biology. In fresh, predominantly stagnant waters, normally in small acid pools with some vegetation, often in *Sphagnum*-pools or in shallow waters on moors, also in woodland. Life cycle as *aequalis*. March-October, mainly in spring, frequently found in drift on the seashore.

123

55. *Helophorus obscurus* Mulsant, 1844
Figs 154, 204.

Helophorus obscurus Mulsant, 1844, Hist. Nat. Col. Fr. Palp.: 36.
Helophorus walkeri Sharp, 1916, Entomologist's mon. Mag. 52: 108.

3.0-4.0 mm. Very similar to *flavipes* in colour and size; lateral margins of pronotum normally reddish; very seldom the elytra are pale straw yellow, with distinct dark sutural arrow-mark. Head and pronotum with close granulation, which is reduced and very weak (often indistinct) on clypeus and middle portion of pronotum; very seldom with strong and well defined granulation (as in *asperatus*). Frontal furrow expanded anteriorly. Submedian furrow of pronotum curved outwards in middle, normally stronger than in *flavipes*, and usually distinctly angular (Fig. 154). Elytra a little wider than in *flavipes,* on the average with more bluntly rounded apex; the striae usually a little stronger, the interstices slightly more convex. Pseudepipleura normally narrow and hardly visible from below, only very seldom wider. Antennae 9-segmented. The external characters (of both *flavipes* and *obscurus*) show some variation, and the two species can sometimes be very difficult to separate on the basis of these, so examination of the aedeagus, which provide good distinguishing characters (Fig. 204) can be necessary.

Distribution. Common or rather common in Denmark and southern Sweden (Sk., Bl., Hall., Öl., Gtl., Vg.); not in Norway and Finland. – Europe, ranging from West Europe (including the British Isles) to the Caucasus, from southern Scandinavia to North Spain, Italy, the Balkans and southern USSR. – North and Central European specimens belongs to ssp. *obscurus* Muls.; other subspecies occur in North Africa and on Corsica and Sardinia.

Biology. Mainly in stagnant, neutral or basic water, normally in shallow, eutrophic, grassy, often temporary pools with somewhat clayey bottom, usually in open country. Sometimes also at the grassy edges of slower reaches of fairly soft-bottomed streams. Life cycle as in *aequalis*. March-November, mainly found in spring; often in drift on the seashore.

Note. Kangas (1968b) recorded the species from East Fennoscandia (as *H. walkeri*). According to Angus (1974) this record is very unlikely, and it is possible that Kangas' specimens belong to *H. discrepans*.

SUBFAMILY SPHAERIDIINAE

Body contour evenly curved, not interrupted between pronotum and elytra. Head and pronotum without distinct impressions, surface rather evenly convex (except in a few exotic forms). Mesosternum strongly elevated, its middle portion often forming a well defined plate, which contacts anterior margin of raised middle portion of metasternum. Abdomen with 5 visible sternites. Maxillary palpi distinctly shorter than antennae, their 2nd segment dilated, much thicker than the other segments (Fig. 224). Antennae 8- or 9-segmented, club mostly compact, only seldom loose. Tarsi 5-segmented, basal segment of meso- and metatarsi longer than 2nd segment (except in some exotic genera). Tibiae without fringes of long swimming hairs. In ♂ ♂ of some genera (e.g. *Cercyon, Megasternum, Cryptopleurum*) the maxillae are provided with a peculiar sucking-disc shaped plate (Fig. 224).

A large subfamily, comprising about 700 known species and numerous genera; represented in all parts of the world, mainly in the warmer climates. 6 genera occur in Europe.

Contrary to almost all other Hydrophilids, the majority of the species in this subfamily are terrestrial, in the sense that they are often found far from water. They live mainly in dung or decomposing plant debris, but only where there is a certain degree of humidity.

Another characteristic is that several of the species apparently have two generations a year, whereas virtually all other hydrophiloids have only one generation a year.

Key to genera of Sphaeridiinae

1 Lateral margins of head not abruptly narrowed before eyes, concealing base of antennae. Anterior margin of eyes distinctly emarginate (lateral view) (Figs 205, 208) (Tribe Sphaeridiini). Length 3.8-7.5 mm 2
 - Lateral margins of head abruptly narrowed before eyes, exposing base of antennae. Anterior margin of eyes not emarginate (lateral view) (Fig. 217). Length 1.4-4.2 mm. Elytra punctato-striate 4
2 (1) Elytra with regular striae or series of punctures. Antennae 9-segmented, club compact . . . . *Dactylosternum* Wollaston (p. 128)
 - Elytra (besides sutural stria) without distinct striae or series of punctures (or at most with obsolete and incomplete series). If antennae 9-segmented, the club is loose 3
3 (2) Antennae 9-segmented, club loose (Fig. 206). Scutellum about as long as wide. Elytra black, without paler spots . .
 . . . . *Coelostoma* Brullé (p. 126)
 - Antennae 8-segmented, club compact (Fig. 209) Scutellum markedly longer than wide. Elytra black, usually with well marked yellowish or reddish spots . . . . *Sphaeridium* Fabricius (p. 129)
4 (1) Elytral epipleura broadly visible, at least in anterior third (Fig. 221). Prosternum ridged in middle, at least in posterior half, at most with a small emargination posteriorly (Figs 219, 220) (Tribe Cercyonini). Dorsal surface without distinct pubescence. Outer face of protibiae without distal excision (except in *C. littoralis*: 2.5-3.3 mm) . . . . *Cercyon* Leach (p. 134)

\-   Elytral epipleura very narrow, except at extreme base (Fig. 260). Prosternum raised in middle to form a well defined flat plate, markedly emarginate posteriorly (Fig. 262) (Tribe Megasternini). If dorsal surface is glabrous, the outer face of protibiae is markedly excised distally (Fig. 257), and the size not exceeding 2.0 mm 5

5 (4) Dorsal surface with distinct pubescence. Outer face of protibiae without distal excision .... *Cryptopleurum* Mulsant (p. 161)

\-   Dorsal surface glabrous. Outer face of protibiae markedly excised distally (Fig. 257) .... *Megasternum* Mulsant (p. 158)

Genus *Coelostoma* Brullé, 1835

Coelostoma Brullé, 1835, Hist. Nat. Ins. 5 (2): 293.

   Type species: *Hydrophilus orbicularis* Fabricius, 1775, by monotypy.

A genus of medium-sized, short and wide, highly and evenly convex species. Eyes anteriorly distinctly emarginate (lateral view). Pronotum short and wide, widest at base, the posterior margin straight. Scutellum about as long as wide. Elytra with irregular, very uniform punctuation and with a fine sharply impressed sutural stria, reaching from apex anteriorly at least to middle; otherwise without striae. Ventral surface of body rather dull, finely shagreened and with dense hydrofuge pubescence. Prosternum normally not ridged; mesosternum with a median longitudinal keel, which is strongly raised posteriorly to form a somewhat rhombic process; metasternum markedly and rather narrowly raised in middle, greatly prolonged anteriorly between mesocoxae, anterior margin closely articulating to the mesosternal process; raised middle portion of metasternum glabrous and shining. Abdominal sternites not ridged medially. Antennae (Fig. 206) 9-segmented, with loose 3-segmented club. Legs rather short, femora and tibiae fairly stout, tarsi much thinner.

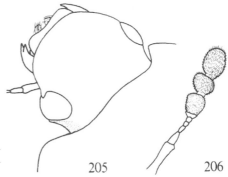

Figs 205, 206. *Coelostoma orbiculare* (F.). – 205: head in dorso-lateral view; 206: right antenna.

A large genus, comprising about 100 known species, represented in all parts of the Old World, mainly in the tropics. Only one species occurs in North Europe.

56. *Coelostoma orbiculare* (Fabricius, 1775)
    Figs 205-207; pl. 2: 9.

Hydrophilus orbicularis Fabricius, 1775, Syst. Ent.: 229.

4.0-4.8 mm. Black; pronotal margins, particularly posterior margin, sometimes narrowly reddish. Entire dorsal surface rather shining, with rather fine and dense punctuation, between the punctures without microsculpture. Maxillary palpi piceous to black, antennae reddish with blackish club, legs dark reddish brown, tarsi slightly paler.

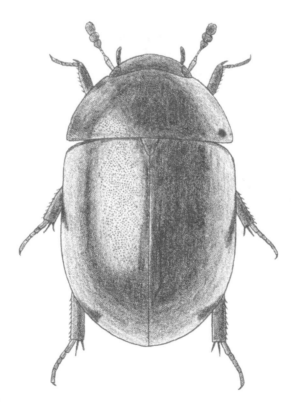

Fig. 207. *Coelostoma orbiculare* (F.), length 4.0-4.8 mm. (After Victor Hansen).

127

Distribution. Widespread, common in Denmark, entire Sweden (except Lapland); widespread in most of southern Norway north to NSy; South and Central Finland north to Ks. – Palearctic; widely distributed in Europe, from Spain and Great Britain to northern Asia and Japan, southwards to the Mediterranean and Asia Minor.

Biology. In stagnant fresh water, mainly at the edges of eutrophic, open, well vegetated ponds, among vegetation in shallow water at the edge. Normally a typical fen-land species, but also in acid or less eutrophic waters, e.g. in *Sphagnum*-bogs. Egg cocoons are found in May, full grown larvae in August-September. The adults are found from March to October, most frequently in spring. Sometimes found in drift on the seashore.

Genus *Dactylosternum* Wollaston, 1854

Dactylosternum Wollaston, 1854, Ins. Mad.: 99.
   Type species: *Dactylosternum rousseti* Wollaston, 1854 (= *Sphaeridium abdominale* Fabricius, 1792), by monotypy.

Body shape similar to *Coelostoma,* but generally less convex, and with less rounded sides, elytra in anterior half subparallel. Eyes anteriorly with a deep, rather narrow emargination (lateral view). Posterior margin of pronotum almost straight. Elytra with 11 regular striae or series of punctures (i.e. besides the usual 10 striae with an extra marginal series). Ventral surface of body dull, very finely shagreened and with dense hydrofuge pubescence. Prosternum (at least anteriorly) ridged in middle, tectiform; mesosternum with a median longitudinal keel, which is strongly raised and enlarged posteriorly, forming a rhombic process; metasternum with shining raised middle portion, which is greatly, rather narrowly prolonged anteriorly between mesocoxae, closely articulating with mesosternal process; the raised metasternal middle wider and less strongly elevated than in *Coelostoma.* 1st abdominal sternite with a well marked median longitudinal ridge. Antennae 9-segmented with compact 3-segmented club. Legs rather short, femora and tibiae rather stout, flattened dorsoventrally, tarsi much thinner.
   The genus comprises about 70 known species represented in all major zoogeographical regions, yet mainly in the tropics. A single species occurs in Europe.

Dactylosternum abdominale (Fabricius, 1792)

Sphaeridium abdominale Fabricius, 1792, Ent. Syst. 1 (1): 79.
Coelostoma insulare Laporte de Castelnau, 1840, Hist. Nat. An. Art. 2: 59.
Dactylosternum rousseti Wollaston, 1854, Ins. Mad.: 100.

3.8-5.0 mm. Black, pronotal margins sometimes narrowly reddish; ventral surface of body reddish brown to piceous brown. Entire dorsal surface rather shining, finely and closely punctuated. Pronotum laterally with extremely fine microsculpture of mostly

128

irregular, oblique waves; dorsal surface otherwise without microsculpture. Elytra with closely punctured striae, the punctures becoming gradually weaker and less close anteriorly (not quite reaching elytral base); the striae laterally more close-set and stronger punctured. Maxillary palpi and antennae yellowish red, antennal club (and sometimes basal segment) slightly darker, legs dark reddish brown.

Distribution. So far not recorded from Scandinavia. – Widely distributed throughout the tropics of both the Old and New World, and in the adjacent temperate zones; thus occurring in South Europe (the Mediterranean), and occasionally found (introduced) in Central Europe as far north as Holstein (Hamburg). Possibly the species may also be found in South Scandinavia.

Biology. A very euryoecious species, in all kinds of decaying organic matter; in Central Europe mainly found in rotting vegetables, but otherwise it is taken in different kinds of refuse (e.g. empty shells of crabs etc., or decaying garden refuse), in manure, etc. Apparently the species easily adapts to very different conditions, and shows a pronounced ability of dispersal, though this is to some extent effected by man. February, April-June, August-October.

Genus *Sphaeridium* Fabricius, 1775

Sphaeridium Fabricius, 1775, Syst. Ent.: 66.
   Type species: *Dermestes scarabaeoides* Linnaeus, 1758, by subsequent designation (Latreille, 1810).
Medium-sized, moderately convex species. Eyes anteriorly with a deep, rather narrow emargination (lateral view). Posterior margin of pronotum feebly bisinuate. Scutellum

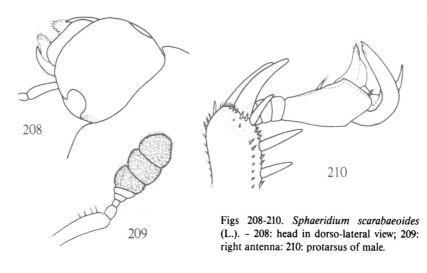

208

209

210

Figs 208-210. *Sphaeridium scarabaeoides* (L.). – 208: head in dorso-lateral view; 209: right antenna: 210: protarsus of male.

elongate, about twice as long as wide. Elytra irregularly, uniformly punctuated, with a fine sharp sutural stria, reaching from apex, anteriorly at least to middle; elytral apices separately rounded. Ventral surface of body dull, very finely shagreened and with dense hydrofuge pubescence. Prosternum bluntly elevated in middle, posteriorly between procoxae prolonged into a rather long sharp process bearing a strong apical spine; mesosternum with a wide, strongly raised obtuse median longitudinal keel; metasternum with a comparatively wide, shining raised middle portion, which is hardly prolonged anteriorly between mesocoxae. Abdominal sternites not ridged in middle. Antennae (Fig. 209) 8-segmented with compact 3-segmented club and very long basal segment. Legs, especially the tarsi, longer than in *Coelostoma,* femora and tibiae fairly thick, flattened dorso-ventrally, tibiae provided with long and strong spines. Protarsi in ♂, especially the claw segment, strongly enlarged, the exterior claw strongly swollen and basally strongly curved (Fig. 210).

A genus of about 30 known species, represented in all the major zoogeographical regions, mainly in warmer climates (missing only in the Australian region). 3 species occur in North Europe.

Key to species of *Sphaeridium*

1   Posterior margin of pronotum distinctly bisinuate, posterior angles sharp (Fig. 215). Common apical elytral spot

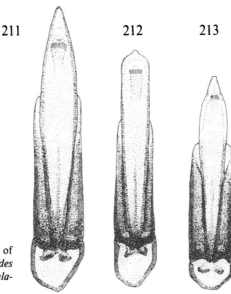

211   212   213

Figs 211-213. Male genitalia of *Sphaeridium.* – 211: *scarabaeoides* (L.); 212: *lunatum* F.; 213: *bipustulatum* F.

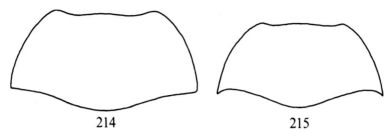

214   215

Figs 214, 215. Pronotum of *Sphaeridium*. – 214: *scarabaeoides* (L.); 215: *bipustulatum* F.

(if present) not divided by a dark sutural stripe. Length
4.0-5.7 mm .... 57. *bipustulatum* Fabricius
– Posterior margin of pronotum more weakly bisinuate, posterior angles somewhat obtuse (Fig. 214). Common apical elytral spot at least partially divided by a dark narrow
sutural stripe. On the average larger, 5.0-7.5 mm 2
2 (1) Pronotum black with lateral margins (at least anteriorly)
reddish yellow. Elytra with a distinct, well defined, large
subhumeral reddish spot. Meso- and metafemora yellowish red, darkened only in middle and along anterior margin................................... 59. *scarabaeoides* (Linnaeus)
– Pronotum entirely black. Subhumeral spot of elytra darker,
usually rather vaguely demarcated or even indistinct. Meso-
and metafemora almost entirely dark .... 58. *lunatum* Fabricius

57. *Sphaeridium bipustulatum* Fabricius, 1781
Figs 213, 215.

Sphaeridium bipustulatum Fabricius, 1781, Spec. Ins. 1: 78.

4.0-5.7 mm. Black, lateral margins of pronotum and elytra reddish yellow, on elytra
occasionally only partially so. Elytra apically with a common, undivided yellowish
red spot, which is very variable in size and shape, and is often rather vaguely demarcated, or may be absent; each elytron also often with a vague, dark reddish, very variable subhumeral spot, which is normally less distinct than apical spot. Entire dorsal
surface very finely and closely punctured, elytra between the punctures with extremely fine microsculpture, and sometimes also with traces of incomplete longitudinal series of sparse and fine punctures. Posterior margin of pronotum distinctly bisinuate,
posterior angles sharp (Fig. 215). Maxillary palpi and antennae piceous to black, the
latter often brownish towards base; legs yellowish red, femora darkened in middle,
apex and outer and inner face of tibiae (except basally) narrowly darkened.
♂: Aedeagus, Fig. 213.

Distribution. Common in the whole of Denmark, and in Sweden north to Lu.Lpm. and Nb.; in Norway recorded from a number of provinces in the south and from NTi; southern Finland north to Kb. – Widely distributed throughout the Palearctic region and North America (introduced).

Biology. Typically in droppings of various mammals, particularly fresh horse and cow dung, but also in most other kinds of decaying organic matter, e.g. compost heaps, rotting plant debris, old mushrooms, carrion, etc. April-October, full grown larvae are found in September-October. Sometimes in drift on the seashore.

58. *Sphaeridium lunatum* Fabricius, 1792
Fig. 212.

Sphaeridium lunatum Fabricius, 1792, Ent. Syst. 1 (1): 78.

5.5-7.5 mm. Very similar to the following species in shape and size, but darker. Pronotum entirely black, without paler margins. Common apical elytral spot normally not (or only indistinctly) continued anteriorly along lateral margin; subhumeral spot normally much darker than in *scarabaeoides,* more vaguely demarcated and often rather indistinct. Legs, especially femora darker, meso- and metafemora almost entirely dark, only paler at apex; metafemora sometimes also paler basally.
♂: Aedeagus, Fig. 212.

Distribution. Common in entire Denmark, and in Sweden north to Lu.Lpm. and Nb.; in Norway only recorded from Ø, AK, HEs, Os, VE, MRi and STi, but no doubt overlooked; whole Finland north to ObN. – Widely distributed, known from entire Europe, and undoubtedly more widespread in the Palearctic region; also very widespread in North America (introduced).

Biology. More stenoecious than *bipustulatum,* predominantly in droppings of various mammals, typically fresh cow and horse dung, only rarely in other kinds of decaying organic matter. April-October; sometimes found in drift on the seashore.

59. *Sphaeridium scarabaeoides* (Linnaeus, 1758)
Figs 208-211, 214, 216; pl. 3: 1.

Dermestes scarabaeoides Linnaeus, 1758, Syst. Nat. (10) 1: 356.

5.0-7.0 mm. Black, lateral margins of pronotum reddish yellow, sometimes only anteriorly; elytra with a common, variable but normally large yellowish apical spot, divided (at least partially) by a narrow dark sutural stripe, and laterally continued anteriorly along elytral margin, at least to middle; each elytron with a large, oblong, somewhat variable, subhumeral red or reddish spot, which is almost always rather well defined and very distinct. Punctuation of dorsal surface very fine and close. Posterior margin of pronotum only feebly bisinuate, posterior angles somewhat obtuse (Fig. 214). Maxillary palpi and antennae piceous to black, the latter often brownish towards base, legs yellow red, femora darkened in middle and along anterior margin (at least meso- and metafemora yellowish red basally and apically); apex and outer and inner face of tibiae (except basally) normally narrowly darkened.

♂: Aedeagus, Fig. 211.

Distribution. Common and widespread in entire Denmark and Sweden north to Lu.Lpm. and Nb.; in Norway most provinces north to TRy; whole Finland north to ObN and Ks. – Widespread throughout the Palearctic region and North America (introduced).

Biology. As *lunatum;* adults are found in February-September, full grown larvae in June-July.

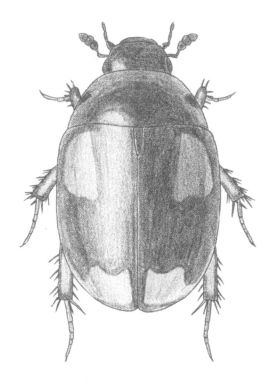

Fig. 216. *Sphaeridium scarabaeoides* (L.), length 5.0-7.0 mm. (After Victor Hansen).

133

Genus *Cercyon* Leach, 1817

Cercyon Leach, 1817, Zool. Misc. 3: 95.

Type species: *Dermestes melanocephalus* Linnaeus, 1758, by subsequent designation (Thomson, 1859).

Dorsal face glabrous, or with extremely fine microscopical hairs. Lateral margins of head rather abruptly narrowed before eyes, not concealing base of antennae; anterior margin of eyes rounded, without emargination (Fig. 217). Posterior margin of pronotum almost straight. Scutellum about equally long as wide. Elytra with 10 finely punctured striae, the lateral striae (about 6th to 10th) usually disappearing a little before reaching base; 10th stria to some extent rudimentary, posteriorly reaching only to middle or a little behind middle, in some species very distinct, in others obsolete. Between striae normally with distinct irregular punctuation, and often also with microsculpture, which may consist of short irregular lines or may form regular isodiametric reticulation. Epipleura well developed, broadly visible at least in anterior third. Ventral surface of body (except raised portions of meso- and metasternum) generally dull, very finely and densely shagreened and with dense hydrofuge pubescence. Prosternum (Figs 219, 220) tectiform, in middle ridged, at most with a small emargination posteriorly, its middle portion (in European species) not differentiated laterally from antennal excavations of prothorax. Mesosternum strongly raised in middle, to form a more or less elongate plate (Fig. 223) which may be very narrow and even strongly ridged. Middle portion of metasternum shining and glabrous, markedly raised to form a more or less well defined pentagonal area, which is hardly prolonged anteriorly between mesocoxae. Metasternum on each side, along postero-lateral margin of raised middle often with a fine oblique line (femoral line) (Fig. 223), which may even be continued to anterior angles of metasternum. 1st abdominal sternite with a well marked longitudinal ridge in middle. Antennae (Fig. 218) 9-segmented with compact 3-segmented club. Legs moderately long, tibiae with comparatively fine spines. Mesocoxae narrowly separated.

A large genus with a world wide distribution, comprising more than 200 known species. 23 species occur in North Europe.

Fig. 217. Head of *Cercyon*. – A: *laminatus* Sharp; B: *impressus* (Sturm).
Fig. 218. Right antenna of *Cercyon depressus* Stph.
Figs 219, 220. Prosternum of *Cercyon*. – 219: *unipunctatus* (L.); 220: *tristis* (Ill.).
Fig. 221. Epipleural portion of elytra of *Cercyon unipunctatus* (L.).
Fig. 222. Protibia of *Cercyon littoralis* (Gyll.).
Fig. 223. Meso- and metasternum of *Cercyon impressus* (Sturm).
Fig. 224. Maxilla of *Cercyon impressus* (Sturm), male.

134

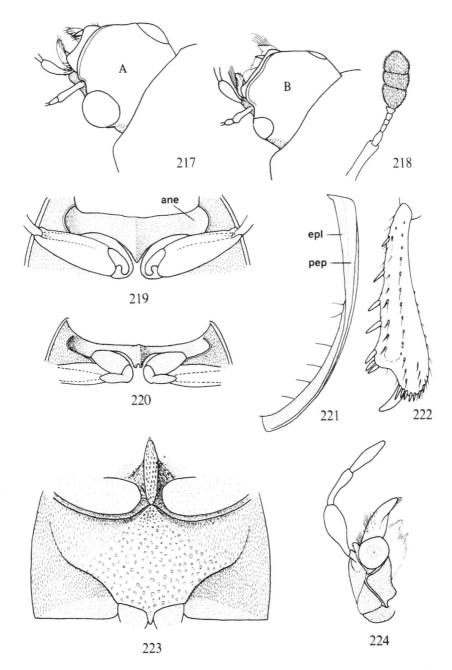

217

A

B

218

ane

219

epl

pep

220

221

222

223

224

Key to species of *Cercyon*

1   Raised middle portion of mesosternum posteriorly broadly contacting metasternum, received in a v-shaped excision at anterior margin of metasternum (Fig. 237). Shining species, black with yellowish to reddish elytral apex. Elytra strongly narrowed posteriorly (Fig. 243). . 82. *analis* (Paykull)

–   Raised middle portion of mesosternum contacting anterior margin of raised metasternal portion in a single point, or separated from it by a narrow gap (Figs 234-236). Elytra less strongly narrowed posteriorly 2

2 (1)   Pronotum strongly convex, more so than elytra, in lateral view not forming a continuous curve with elytra . . 60. *ustulatus* (Preyssler)

–   Pronotum not more convex than elytra; pronotum and elytra in lateral view forming a continuous curve 3

3 (2)   Eyes very large, globular (Fig. 217A). Raised middle portion of mesosternum tectiform, narrow, sharply and highly ridged in entire length. Pronotum and elytra concolorous, rather pale brownish, towards margins often paler. Large species; 3.2-4.0 mm . . . . 61. *laminatus* Sharp

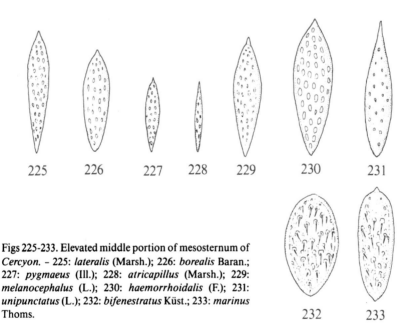

Figs 225-233. Elevated middle portion of mesosternum of *Cercyon*. - 225: *lateralis* (Marsh.); 226: *borealis* Baran.; 227: *pygmaeus* (Ill.); 228: *atricapillus* (Marsh.); 229: *melanocephalus* (L.); 230: *haemorrhoidalis* (F.); 231: *unipunctatus* (L.); 232: *bifenestratus* Küst.; 233: *marinus* Thoms.

136

– Eyes distinctly smaller, less convex (Fig. 217B). Raised middle portion of mesosternum wide or narrow, forming a well defined plate. Pronotum darker, often black (except in *atricapillus:* 1.5-2.0 mm), lateral margins often paler 4

4 (3) Outer face of protibiae in about distal fifth distinctly excised, with a large strong spine distal to the excision (Fig. 222) .... 62. *littoralis* (Gyllenhal)
– Outer face of protibiae without distinct excision 5

5 (4) Lateral margins of pronotum feebly, but distinctly sinuate posteriorly .... 63. *depressus* Stephens
– Lateral margins of pronotum rounded, without sinuation 6

6 (5) Elytra without microsculpture, or with microsculpture consisting of fine irregular lines, not reticulate 7
– Elytra with distinct, though sometimes very fine, reticulate, very regular microsculpture. Short and strongly convex species (as Fig. 242); black, at most elytral apex paler 20

7 (6) Metasternum with distinct femoral lines (Fig. 223)................... 8
– Metasternum without femoral lines 12

8 (7) Smaller, 1.4-2.0 mm. Maxillary palpi reddish, terminal segment often darker. Femoral lines of metasternum continued to anterior angles of metasternum 9
– Larger, 2.3-3.6 mm. Maxillary palpi piceous to black. Femoral lines of metasternum not reaching anterior angles of metasternum (Fig. 223) 10

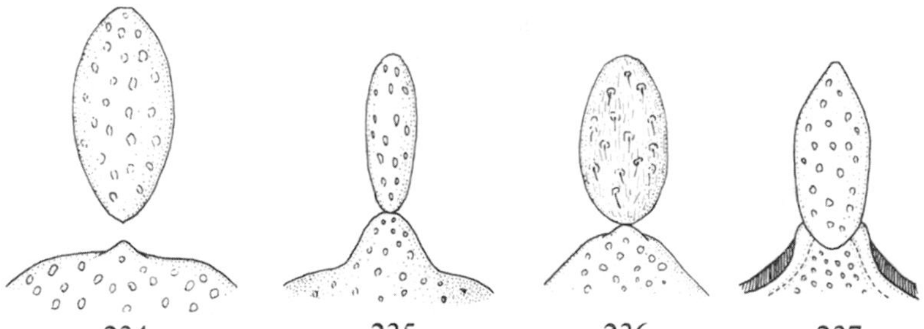

Figs 234-237. Elevated middle portion of mesosternum and anterior part of elevated mid-portion of metasternum in *Cercyon*. - 234: *tristis* (Ill.); 235: *convexiusculus* Stph.; 236: *sternalis* (Sharp); 237: *analis* (Payk.).

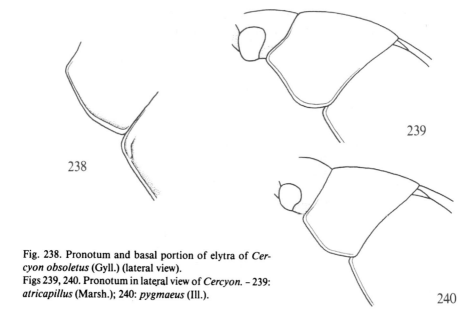

Fig. 238. Pronotum and basal portion of elytra of *Cercyon obsoletus* (Gyll.) (lateral view).
Figs 239, 240. Pronotum in lateral view of *Cercyon*. – 239: *atricapillus* (Marsh.); 240: *pygmaeus* (Ill.).

9 (8) Lateral margins of pronotum very strongly rounded in posterior third (lateral view) (Fig. 239). Innermost elytral striae distinct to base. Moderately convex species ......
.... 77. *atricapillus* (Marsham)
– Lateral margins of pronotum less strongly rounded (lateral view) (Fig. 240). Innermost elytral striae, especially 2nd stria, obsolete anteriorly. More convex species . 76. *pygmaeus* (Illiger)
10 (8) Posterior margin of pronotum with a small pit-like depression in middle. Elytra rather strongly narrowed posteriorly (Fig. 241). Size on the average larger, 3.0-3.6 mm .
.... 65. *impressus* (Sturm)
– Posterior margin of pronotum without pit-like depression. Elytra less strongly narrowed posteriorly. Size on the average smaller, 2.3-3.2 mm 11
11 (10) Elytra red or yellowish red, with a sharply defined triangular black basal band, black anterior angles and black epipleura (Fig. 253) .... 67. *melanocephalus* (Linnaeus)
– Elytra usually more extensively darkened, or if paler, with a black T-shaped (never triangular) basal mark, covering basal elytral margin and anterior sutural portion; epipleura reddish or yellowish brown ..... 66. *haemorrhoidalis* (Fabricius)

12 (7) Large species, 3.5-4.2 mm. Lateral elytral margin just behind anterior angle with a small oblique pit-like depres-
sion (Fig. 238); maxillary palpi piceous to black . 64. *obsoletus* (Gyllenhal)
–   Smaller, 1.6-3.5 mm. Lateral elytral margin without such
pit-like depression. Maxillary palpi mostly paler, often
yellowish red 13
13 (12) Anterior margin of clypeus concave. Elytra (Fig. 254)
black with large yellowish red medio-apical spot, narrowly divided by a dark sutural stripe, and some reddish
sub-basal spots (at least in 6th interstice); apical spot continued anteriorly as a narrow reddish stripe (covering at

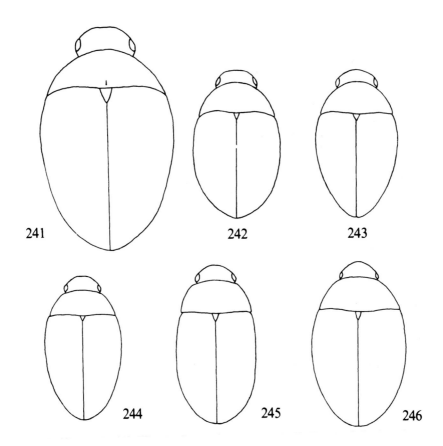

Figs 241-246. Outline of *Cercyon*. – 241: *impressus* (Sturm); 242: *tristis* (Ill.); 243: *analis* (Payk.);
244: *terminatus* (Marsh.); 245: *quisquilius* (L.); 246: *borealis* Baran.

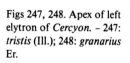

Figs 247, 248. Apex of left elytron of *Cercyon*. – 247: *tristis* (Ill.); 248: *granarius* Er.

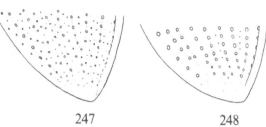

247   248

least 4th interstice), usually connected to sub-basal spots; lateral elytral margins narrowly yellowish red ....
.... 68. *emarginatus* Baranowski
–   Anterior margin of clypeus truncate, not concave. Elytra differently coloured 14

14 (13) Elytra black with sharply defined apical yellow spot, which is often narrowly continued anteriorly along lateral margin (Figs 249, 250). Raised middle portion of mesosternum rather wide (Figs 232, 233) 15
–   Elytra differently coloured, or if black with sharply defined yellow apical spot, this is much broader continued anteriorly along lateral margin (Fig. 251), and raised middle portion of mesosternum is markedly narrower (Fig. 231) 16

15 (14) Yellow apical spot covering about posterior fourth to fifth of elytra, laterally continued anteriorly at most to middle (Fig. 250). Raised middle portion of mesosternum about twice as long as wide (Fig. 232) .... 71. *bifenestratus* Küster
–   Yellow apical spot covering only about posterior eighth or seventh of elytra, but laterally continued further forwards, at least to middle (Fig. 249). Raised middle portion of mesosternum narrower, about 3 times as long as wide (Fig. 233) .... 72. *marinus* Thomson

16 (14) Elytra yellow, in about middle with a large well defined blackish common sutural spot (Fig. 252), which sometimes is greatly enlarged (Fig. 251) so elytra may be black with sharply defined yellow apical spot, which laterally is broadly continued anteriorly to elytral base ....
.... 73. *unipunctatus* (Linnaeus)
–   Elytra differently coloured 17

17 (16) Body shape rather parallel-sided (Fig. 245), elytra rather narrow, uniformly yellow (seldom slightly darker round scutellum). Pronotum black, lateral margins yellowish at least anteriorly........................... 74. *quisquilius* (Linnaeus)
–   Body shape less parallel-sided (Figs 244, 246). Elytra dar-

ker, reddish or brownish, of if almost uniformly pale
(rare specimens of *terminatus*), pronotum without paler
margins 18

18 (17) Pronotum uniformly black. Maxillary palpi uniformly
yellowish red. Smaller, 1.6-2.3 mm . . . . 75. *terminatus* (Marsham)

– Pronotum black with yellowish or reddish lateral margins. Maxillary palpi yellowish red (*lateralis:* at least 2.5
mm) or with darker terminal segment. Normally larger,
2.3-3.2 mm 19

19 (18) Elytra widest in about anterior fourth, normally rather
strongly narrowed posteriorly (almost as in Fig. 241).
10th elytral stria markedly finer than 9th. Maxillary

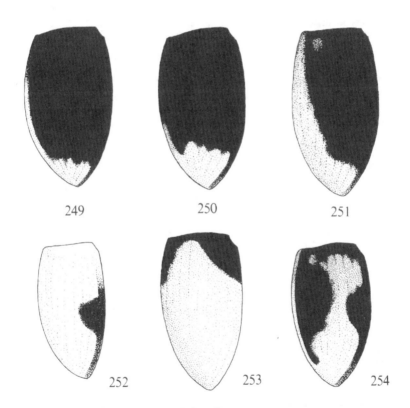

Figs 249-254. Colour pattern of left elytron in *Cercyon*. – 249: *marinus* Thoms.; 250: *bifenestra-tus* Küst.; 251: *unipunctatus* var. *janssoni* Nyholm; 252: *unipunctatus* (L.); 253: *melanocephalus* (L.); 254: *emarginatus* Baranowski.

palpi almost uniformly reddish. Raised middle portion
of mesosternum very narrow (Fig. 225) . . . . 70. *lateralis* (Marsham)

- Elytra a little wider, with slightly more rounded sides,
less strongly narrowed posteriorly (Fig. 246). 10th elytral
stria not distinctly finer than 9th. Terminal segment of
maxillary palpi distinctly darkened. Raised middle por-
tion of mesosternum wider (Fig. 226) . . . . 69. *borealis* Baranowski

20 (6) Raised middle portion of mesosternum posteriorly separated from anterior margin of raised metasternal por-
tion by a narrow gap (Fig. 234) 21

- Raised middle portion of mesosternum posteriorly contacting anterior margin of raised metasternal portion
(Figs 235, 236) 22

21 (20) Elytral striae, particularly the lateral striae, becoming rather obsolete and somewhat irregular well before apex
(Fig. 247). Elytra normally rather dull, with very distinct
microsculpture . . . . 79. *tristis* (Illiger)

- Elytral striae well defined, even apically (Fig. 248). Elytra rather shining, especially laterally with much finer
microsculpture . . . . 78. *granarius* Erichson

22 (20) Raised middle portion of mesosternum wide (Fig. 236).
Maxillary palpi almost uniformly reddish. Punctuation
of elytral interstices extremely fine, often indistinct, espe-
cially laterally and posteriorly . . . . 81. *sternalis* (Sharp)

- Raised middle portion of mesosternum narrower (Fig.
235). Terminal segment of maxillary palpi darkened.
Punctuation of elytral interstices stronger, also very dis-
tinct laterally and posteriorly . . . . 80. *convexiusculus* Stephens

Subgenus *Dicyrtocercyon* Ganglbauer, 1904

Dicyrtocercyon Ganglbauer, 1904, Käf. Mit. 4: 274.
Type species: *Sphaeridium ustulatum* Preyssler, 1790, by monotypy.

Pronotum strongly convex, longitudinally more convex than elytra, in lateral view not
forming a continuous curve with elytra. Raised middle portion of mesosternum forming a rather narrow elongate plate, which is posteriorly separated from anterior margin
of raised metasternal portion by a narrow gap.

60. *Cercyon ustulatus* (Preyssler, 1790)

Sphaeridium ustulatum Preyssler, 1790, Verz. böhm. Ins.: 34.

2.6-3.4 mm. A short and wide, strongly convex species, easily recognized by the highly arched pronotum. Black; elytra with a large red or yellowish red apical spot, which is narrowly divided medially by a dark sutural stripe. Head and pronotum finely and densely punctured. Anterior margin of clypeus truncate, not concave. Elytral striae fine, laterally and posteriorly a little stronger; lateral striae with fine, distinct punctures; inner striae more indistinctly punctured. Interstices shining, anteriorly almost flat, very finely punctured and without distinct microsculpture; interstices laterally and posteriorly becoming gradually more convex and more finely punctured. Raised middle portion of mesosternum narrow (index length: width about 4). Metasternum without femoral lines. Maxillary palpi and antennae reddish, the latter with darker club; legs reddish brown or brown, tarsi paler.

Distribution. Widespread and not uncommon in South and Central Scandinavia; found in most parts of Denmark; Sweden north to Vstm. and Vrm.; in Norway only Ø and AK; southern Finland north to Sb.. – North and Central European species, south to the northern Mediterranean, ranging from West Europe to Siberia; also in eastern North America (introduced).

Biology. On muddy banks of stagnant or running waters, mainly in wet mud near the edge of the water or under moss, leaf litter, pieces of wood, etc. January, March-October.

Subgenus *Paracycreon* d'Orchymont, 1942

Paracycreon d'Orchymont, 1942a, Bull. Mus. r. Hist. nat. Belg. 18 (26): 3.
    Type species: *Cercyon hova* Régimbart, 1903, by original designation.

Pronotum not strongly convex, in lateral view forming a continuous curve with elytra. Raised middle portion of mesosternum narrow, sharply and highly ridged in entire length, not forming a distinct plate.

61. *Cercyon laminatus* Sharp, 1873
    Fig. 217A; pl. 3: 7.

Cercyon laminatus Sharp, 1873, Trans. R. ent. Soc. Lond.: 66.

3.2-4.0 mm. Brown or yellowish brown with piceous to black head. Pronotal margins often slightly paler; elytra with paler lateral margins and apex, often also with paler suture and basal margin. Ventral face of body piceous to black, with raised middle portion of metasternum and posterior margins of abdominal sternites yellowish brown. Head and pronotum densely and rather finely punctured. Anterior margin of clypeus truncate, not concave. Eyes larger than in our other species (Fig. 217A), globularly protruding. Elytral striae fine, with very distinct punctures; the striae laterally and posteriorly distinctly deeper impressed, more finely punctured posteriorly. Interstices flat anteriorly, becoming distinctly convex laterally and posteriorly, without distinct

143

microsculpture, densely and finely punctured. Metasternum without femoral lines. Appendages yellowish red, antennal club darker.

Distribution. A newly immigrated species, probably dispersing in Scandinavia; from Denmark still only few findings from F, LFM, SZ, and NEZ; in Sweden as yet only Sk., Gtl. and Nb.; in Finland recorded from more provinces (Ab, N, Ka, St, Ta, Sa, Om); not known from Norway. – Originally described from Japan; since the 1950's found in a number of countries in North and Central Europe (south to North Italy), where in recent decades it has been, and probably still is, in process of dispersal.

Biology. In Central and North Europe the species is found predominantly at light or in drift on the seashore. Obviously it is a very active flyer (flying mainly towards evening), and records from seashores are probably accidental, normally only in connection with periods of windy weather (when the wind is inshore). More likely its natural habitat is among decaying organic matter, such a plant debris; in North Sweden in compost heaps and cow dung. May-September, most abundantly found in the spring.

Subgenus *Cercyon* Leach, 1817, s.str.

Cercyon Leach, 1817, Zool. Misc. 3: 95.
> Type species: *Dermestes melanocephalus* Linnaeus, 1758, by subsequent designation (Thomson, 1859).

(Other subgeneric names, e.g. *Ercycon* Rey, 1886, *Cerycon* Rey, 1886, and *Paraliocercyon* Ganglbauer, 1904, are superfluous).

Pronotum not strongly convex, in lateral view forming a continuous curve with elytra. Raised middle portion of mesosternum forming a more or less narrow, elongate plate, that posteriorly contacts anterior margin of raised metasternal portion, or is separated from it by a narrow gap (Figs 234-236).

62. *Cercyon littoralis* (Gyllenhal, 1808)
Fig. 222; pl. 3: 2.

Sphaeridium littoralis Gyllenhal, 1808, Ins. Suec. 1 (1): 111.
Cercyon litoralis auctt.

2.5-3.3 mm. Brownish to black, elytra paler apically, and often also along lateral margins; very seldom the elytra are reddish or pale yellow, each with a small vague darker spot behind middle near suture. Head and pronotum finely and moderately densely punctured. Anterior margin of clypeus feebly concave. Pronotum in about posterior half subparallel, almost straight sided, at base slightly narrower than base of elytra. Elytra about 1/3 longer than wide, moderately convex, the sides usually rather weakly curved. Elytral striae fine, distinct to apex, anteriorly with fine strial punctures, posteriorly without distinct punctures; strial punctures slightly coarser laterally; 10th stria not much finer than 9th, often rather inconspicuous. Interstices feebly convex or almost flat, rather shining, also apically, with very fine and moderately dense punctuation, and fine microsculpture consisting of sparse, irregular lines. Raised middle portion of mesosternum narrow (index length: width about 4). Metasternum without femoral lines. Maxillary palpi and antennae yellowish red, antennal club darker, legs reddish brown, often rather dark. Outer face of protibiae in about distal fifth distinctly excised, with a large strong spine distal to the excision (Fig. 222) (in all our other species without such an excision).

Distribution. Common and widespread along the seashores of entire Denmark, and in Sweden north to Upl., in southern Finland north to St; in Norway from Ø to Fv. – A holarctic species; in the Palearctic region ranging from Fennoscandia and the Baltic region of the USSR to the Atlantic coasts of West Europe (including Great Britain); southwards to Portugal and the Mediterranean coasts of France and Italy; also recorded from saline inland localities of Great Britain, Rumania and the USSR.

Biology. A halobiontic species, confined to sandy, often somewhat clay-mixed seashores, where it lives in decaying organic matter, predominantly under seaweed, normally very abundant; farther south also found on saline inland localities. Full grown larvae are found at the beginning of June, and also – together with pupae and adults – in August; probably the species has two generations per year. Adults are found especially in spring and autumn.

63. *Cercyon depressus* Stephens, 1829
  Fig. 218.

Cercyon depressum Stephens, 1829, Ill. Brit. Ent. Mandib. 2: 138.

2.2-2.6 mm. Very similar to *littoralis* in shape and colour, but generally a little less convex, and a little more parallel-sided; head and pronotum on the average with slightly denser punctuation. Anterior margin of clypeus truncate, not concave. Pronotal sides distinctly sinuate posteriorly. Elytral striae finer than in *littoralis,* obsolete anteriorly and posteriorly, disappearing well before apex; also lateral striae rather obsolete; 10th stria not distinct. Interstices (at least posteriorly) somewhat dull, punctuation finer than in *littoralis,* or even indistinct, the microsculpture consisting of fine and rather sparse irregular lines, which are however more numerous than in *littoralis,* and posteriorly changes gradually to fine and dense, irregular and somewhat rudimentary reticulation. Meso- and metasternum as in *littoralis.* Protibiae without apical excision, and with smaller apical spine.

Distribution. Widespread at the seashores of Denmark and southern Sweden north to Upl. and Hls.; widespread in Norway north to Nnv; southern Finland. In general not as frequent as *littoralis.* – Europe; ranges from Fennoscandia and the Baltic region of USSR to the Atlantic coasts of West Europe; south to the Mediterranean coasts of France and Italy; also in North America (introduced).

145

Biology. As *littoralis,* and often occurring together with it, but normally much less abundant.

64. *Cercyon obsoletus* (Gyllenhal, 1808)
   Fig. 238.

Sphaeridium obsoletum Gyllenhal, 1808, Ins. Suec. 1: 107.
Sphaeridium lugubris Olivier, 1790, Ent. 2 (15): 7 (mis-interpretation of *Dermestes lugubris* Fourcroy, 1785).

3.5-4.2 mm. Black, elytra usually piceous with distinctly paler (reddish) apex; pale colouring of elytra seldom more extensive. Head and pronotum with fine and dense punctuation. Anterior margin of clypeus truncate, not concave. Pronotum anteriorly strongly narrowed, posterior margin (almost always) without distinct pit-like depression in middle. Elytra rather wide, moderately strongly convex; the striae very fine, inner striae often somewhat obsolete anteriorly; 4th and 5th striae distinctly diverging basally; 10th stria finer than 9th, often indistinct. Interstices flat or almost so; their punctuation slightly finer and denser than on head and pronotum. Elytra at lateral margin, just behind anterior angle with a small oblique pit-like depression (Fig. 238). Raised middle portion of mesosternum narrow (index length: width about 4). Metasternum without femoral lines. Maxillary palpi piceous to almost black, antennae yellowish brown with darker club, legs reddish to brown, tarsi paler.

Distribution. A widely distributed, but in Scandinavia rather uncommon and local species; in Denmark especially in the eastern parts; in Sweden recorded from most of the southern provinces north to Dlr., and from Med., Ås. Lpm. and Nb.; in Norway only AK and Ry; southernmost parts of Finland (Al, Ab, N, Ka, Sa). – Widespread in Europe, though less frequent towards the north; also in North Africa (Algier) and the Caucasus.

Biology. Particularly in droppings of various mammals, e.g. in deer and cow dung, but also in other kinds of decaying organic matter, such as compost heaps, rotting vegetables, old mushrooms, carrion etc., both on open, sunny ground and in woodland, yet avoiding the more shaded sites. May-October, mainly in spring and autumn.

65. *Cercyon impressus* (Sturm, 1807)
   Figs 217B, 223, 224, 241.

Sphaeridium impressum Sturm, 1807, Deutschl. Ins. 2: 9.

3.0-3.6 mm. Black; elytra piceous to black with reddish apex, sometimes paler with the reddish colour extending further forwards so only a large triangular basal band may be black (only rarely with uniformly reddish elytra). Punctuation of head and pronotum almost as in *obsoletus.* Anterior margin of clypeus slightly concave. Posterior margin of pronotum with a small pit-like depression in middle. Elytra more convex than in *obsoletus,* more strongly narrowed posteriorly (Fig. 241); the striae stronger than in *obsoletus;* 4th and 5th striae only slightly diverging basally; 10th stria finer than 9th. Interstices without distinct microsculpture, a little finer and less densely punctured than in *obsoletus.* Lateral margins of elytra without pit-like depression just behind anterior angle. Raised middle portion of mesosternum almost as in *obsoletus* (Fig. 223). Metasternum with femoral lines not quite reaching anterior angles of metasternum (Fig. 223). Maxillary palpi piceous to black, antennae yellowish brown with darker club; legs reddish, tarsi often paler.

Distribution. Very widespread and rather common, found in almost entire Denmark and Fennoscandia. – Widely distributed in North and Central Europe, southwards to the northern Mediterranean, ranging from Great Britain to the European part of the USSR; also in North America (introduced).

Biology. In all kinds of decaying organic matter, but mainly in dung and decaying plant debris, both in open country and in woodland; the species is apparently most abundant in mountainous areas. An active flyer. Found mainly in spring and autumn.

66. *Cercyon haemorrhoidalis* (Fabricius, 1775)
   Fig. 230.

Sphaeridium haemorrhoidalis Fabricius, 1775, Syst. Ent.: 67.

2.5-3.2 mm. Black; elytra usually brownish or reddish with paler apex and lateral margins, and (often) a vaguely paler transverse spot near base; epipleura reddish or yellowish brown. Elytra sometimes rather pale, so dark colouring may be reduced to a black T-shaped mark, covering basal margin and anterior sutural portion. Punctuation on head and pronotum as in *impressus.* Anterior margin of clypeus truncate, very feebly concave. Posterior margin of pronotum without pit-like depression in middle. Elytra moderately convex, striae almost as in *impressus,* though on the average a little finer; 10th stria a little finer than 9th. Interstices flat or almost so, without distinct microsculpture; punctuation anteriorly almost as on pronotum, posteriorly distinctly finer. Raised middle portion of mesosternum rather narrow (index length: width about 3.5) (Fig. 230). Metasternum with femoral lines, not quite reaching anterior angles of metasternum. Maxillary palpi and antennae piceous to black, the latter sometimes slightly paler; legs reddish brown, tarsi paler.

Distribution. Very widespread and common, found in entire Denmark and Fennoscandia. – Widespread throughout the Palearctic region; also in North America (introduced).

Biology. Very euryoecious, found in all kinds of decaying organic matter, mainly in cow, horse and sheep dung, but also frequently in rotting plant debris (particularly compost heaps), old mushrooms, flood-debris (on more wet habitats), carrion, at sap (e.g. birch), etc.; it is also found in nests of various birds. The species is an active flyer, taken in drift on the seashore and at light; it frequently flies also in daytime. April-October.

67. *Cercyon melanocephalus* (Linnaeus, 1758)
Figs 229, 253; pl. 3: 3.

Dermestes melanocephalus Linnaeus, 1758, Syst. Nat. (10) 1: 356.

2.3-3.0 mm. Black; elytra (Fig. 253) red or yellowish red, with a large triangular, black basal band, that reaches laterally to the anterior angles and anterior portion of the lateral margins; epipleura also black. The species seems to be quite constant in regard to this colour pattern. Otherwise it is very similar to *haemorrhoidalis,* but the punctuation of the elytral interstices are finer posteriorly (thus much finer than anteriorly), and raised middle portion of mesosternum a little narrower (index length: width about 4) (Fig. 229). Metasternal femoral lines as in *haemorrhoidalis.* Maxillary palpi and antennae piceous to black, legs reddish brown, often rather dark, tarsi paler.

Distribution. A very common species, found throughout Denmark and Fennoscandia. – Widely distributed; entire Palearctic region. – Records of the species from North America are based on misidentifications.

Biology. Mainly in droppings of various mammals, typically in sheep dung, but also abundant in fresh cow and horse dung, only less frequently in other kinds of decaying organic matter. The species is a very active flyer, and frequently flies also in daytime. March-October.

68. *Cercyon emarginatus* Baranowski, 1985
Fig. 254.

Cercyon emarginatus Baranowski, 1985, Ent. Scand. 15: 344.

2.3-2.8 mm. Piceous to black; lateral margins of pronotum yellowish red; elytra (Fig. 254) piceous to black with a large or very large yellowish red medio-apical spot narrowly divided by a dark sutural stripe; anteriorly with some smaller reddish spots, consisting of a small subhumeral spot near base of 6th interstice, and usually also a larger sub-basal spot (mostly covering interstice 2-4); sub-basal spots often mutually fused; apical spot almost always continued to the anterior as a narrow reddish stripe (covering at least 4th interstice) and usually connected to the sub-basal spot(s); lateral elytral margins narrowly yellowish red. The pale elytral colour is often rather extensive, and usually rather sharply defined. Head and pronotum finely and not very densely punctured, shining. Anterior margin of clypeus feebly, but distinctly concave. Elytral striae fine, only slightly impressed, very finely punctured; 10th stria rather distinct, almost as strong as 9th. Interstices flat or almost so, with somewhat sparse and very fine, or sometimes indistinct punctuation, and with distinct microsculpture consisting of fine irregular lines, which may be rather numerous. Raised middle portion of mesosternum rather narrow (index length: width about 3). Metasternum without femoral lines. Appendages yellowish red; antennal club and terminal segment of maxillary palpi slightly darker.

Distribution. North and Central Fennoscandia; in Sweden found in Vrm., Dlr.,

Äng., Vb., Nb. and P. Lpm., in Norway so far known only from HEs (envir. lake Ut-gardsjøen, 30 km SW of Kongsvinger); in Finland so far only in Ks (Kuusamo); not in Denmark. – Not recorded outside Fennoscandia.

Biology. The species is predominantly found in droppings of mammals, typically in elk dung, but sometimes also at sap of birch stumps. May-September, perhaps mainly in spring and autumn.

69. *Cercyon borealis* Baranowski, 1985
Figs ?26, 246.

Cercyon borealis Baranowski, 1985, Ent. Scand. 15: 341.

2.3-2.8 mm. Similar to *emarginatus* in size and shape (Fig. 246). Piceous to black; lateral margins of pronotum yellowish red; elytra dark reddish, paler apically, and often also slightly paler antero-medially; sometimes however rather light with only in-distinct darkening in middle. Punctuation of head and pronotum as in *emarginatus*. Anterior margin of clypeus truncate, not concave. Elytral striae a little stronger punc-tured than in *emarginatus;* 10th stria conspicuous, not distinctly finer than 9th. Punc-tuation of interstices rather fine, but much stronger than in *emarginatus,* anteriorly usually distinctly stronger and denser than on pronotum, posteriorly and laterally slightly finer. Raised middle portion of mesosternum slightly narrower than in *emar-ginatus* (index length: width about 3.5) (Fig. 226). Metasternum without femoral lines. Appendages yellowish red, antennal club and terminal segment of maxillary palpi dar-ker.

Distribution. North and Central Fennoscandia; in Sweden widespread in the north (Äng., Vb., Nb., Ly. Lpm., P. Lpm., Lu. Lpm.) and in Dlr. and Hls.; in Finland so far only in Kb (Ilomantsi) and Ks (Kuusamo); in Norway so far only Fø (Kirkenes); not in Denmark. – So far not recorded outside Fennoscandia.

Biology. As *emarginatus,* and sometimes found together with it. May-September, found mainly in the summer.

70. *Cercyon lateralis* (Marsham, 1802)
Figs 225, 255.

Dermestes lateralis Marsham, 1802, Ent. Brit. 1: 69.

2.5-3.2 mm. Very similar to *borealis* and of the same colour, except for the maxillary palpi, which are almost uniformly reddish. Punctuation of head and pronotum on the average slightly denser than in *borealis,* and pronotum a little wider. Anterior margin of clypeus truncate, not concave. Elytra widest in about anterior quarter, more strong-ly narrowed posteriorly and normally in more straight lines than in *borealis* (thus in elytral shape somewhat similar to *impressus,* Fig. 241). Elytral striae and interstices almost as in *borealis,* interstices however on the average a little denser punctured, and without distinct microsculpture; 10th stria, though usually distinct, always markedly finer than 9th. Raised middle portion of mesosternum very narrow (index length: width about 5 or more) (Fig. 225). Metasternum without femoral lines; lateral portions of metasternum with some fine and inconspicuous, shallow pit-like punctures.

Distribution. Very common in entire Denmark and Fennoscandia. – Widely distributed in North and Central Europe, southwards to the Mediterranean, eastwards to Siberia; also widespread in North America (introduced).

Biology. Very euryoecious, in all kinds of decaying organic matter, mainly in dung and compost heaps, but also at sap of deciduous trees, in old mushrooms, carrion, etc. It is a very active flyer, often taken at light or in drift on the seashore; frequently flying also at daytime. April-October.

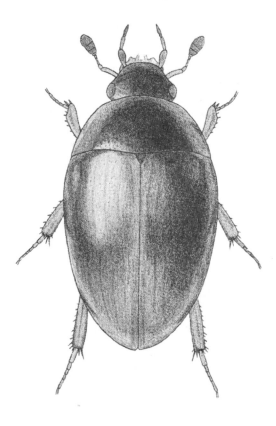

Fig. 255. *Cercyon lateralis* (Marsh.), length 2.5-3.2 mm.

71. *Cercyon bifenestratus* Küster, 1851
   Figs 232, 250.

Cercyon bifenestratum Küster, 1851, Käf. Eur. (23): 15.

2.3-2.8 mm. A rather short and convex species. Black; elytra with a sharply defined large yellow apical spot, medially divided by a narrow dark sutural stripe; apical spot covering about posterior fourth to fifth of elytra, laterally rather narrowly prolonged anteriorly along elytral margin, at most reaching to middle (Fig. 250). Anterior angles of pronotum yellowish. Head and pronotum rather densely and not very finely punctured. Anterior margin of clypeus truncate, not concave. Elytral striae very fine, with very fine punctures, that are only a little coarser than the punctures of the interstices; 10th stria distinctly finer than 9th, sometimes obsolete. Interstices anteriorly a little denser punctured than on pronotum, and with fine or indistinct microsculpture, consisting of very fine irregular lines; the punctuation becoming gradually much finer apically. Raised middle portion of mesosternum rather wide, only about twice as long as wide (Fig. 232), the surface often feebly concave. Metasternum without femoral lines. Appendages reddish brown or brown, usually rather dark, only antennal base paler.

Distribution. Widely distributed, but comparatively local; whole Denmark; scattered records from most Swedish provinces north to Nb. (not in Lapland); in Finland north to ObS; not mentioned from Norway. – North and Central European species, generally local or rare; southwards to northernmost Italy.

Biology. A very stenoecious species, obviously confined to bare or very sparsely covered, sandy or silty, moist banks of fresh water; it is a typical inhabitant of gravel pits, where it lives at the edges of ground water pools. It is a very active flyer, often taken at light or in drift on the seashore. April-June, August, November.

72. *Cercyon marinus* Thomson, 1853
   Figs 233, 249; pl. 3: 4.

Cercyon marinum Thomson, 1853, Öfvers. K. VetenskAkad. Förh. 10: 54.

2.5-3.4 mm. Very similar to *bifenestratus,* but yellow apical spot of elytra smaller, covering only about posterior seventh to eighth of elytra, but along lateral margins prolonged further forwards (at least to middle) (Fig. 249). Punctuation of head, pronotum and elytra as in *bifenestratus,* though elytral punctuation becomes only somewhat finer posteriorly. Raised middle portion of mesosternum rather narrow (index length: width about 3) (Fig. 233), its surface hardly concave. Metasternum without femoral lines. Appendages reddish to brownish; maxillary palpi, especially terminal segment, and antennal club darker.

Distribution. Fairly common; entire Denmark; entire Sweden (with the exception of some northern provinces); in Finland found in most provinces north to ObN; in Norway scattered records from Ø to Fi. – A holarctic species, widespread throughout most of North and Central Europe and northern Asia, eastwards to Japan; widespread also in North America, particularly in the northern regions.

Biology. A markedly hygrophilous species, predominantly living at the edges of fresh, both stagnant and running, water, normally on more well established biotopes than *bifenestratus*. It is usually found in very wet mud, in wet moss (e. g. *Sphagnum*), but sometimes also among decomposing plant debris or other kinds of decaying organic matter. It is an active flyer, often taken at light or in drift on the seashore. March-October.

73. *Cercyon unipunctatus* (Linnaeus, 1758)
   Figs 219, 221, 231, 251, 252; pl. 3: 8.

Coccinella unipunctata Linnaeus, 1758, Syst. Nat. (10)1: 364.
Cercyon janssoni Nyholm, 1952, Ent. Tidskr. 73: 207.

2.4-3.4 mm. Piceous to black; elytra and lateral margins of pronotum yellowish, normally pale; elytra in about middle with a variably large blackish common spot across the suture (Fig. 252) and (at least posteriorly) with a narrow darker sutural stripe. The elytral spot is only extremely rarely indistinct, but may in contrast be greatly enlarged, so the elytra become black, with pale subhumeral spot, and well marked yellow apical spot, which is broadly continued anteriorly along lateral margins, reaching to base (Fig. 251); such specimens (described as *janssoni)* may thus resemble *marinus* and *bifenestratus,* but are recognized by the always more distributed yellow colour, and narrower mesosternal plate. Head and pronotum very densely and rather finely punctured. Anterior margin of clypeus truncate, not concave. Elytral striae fine, with fine, but distinct punctures, which become coarser laterally; 10th stria very conspicuous, not finer than 9th. Elytral interstices flat or almost so, with very fine microsculpture consisting of very fine irregular lines; punctuation anteriorly almost as on pronotum, becoming gradually finer posteriorly. Raised middle portion of mesosternum very narrow (index length: width about 5) (Fig. 231). Metasternum without femoral lines. Appendages reddish, antennal club and terminal segment of maxillary palpi darker (often very dark); appendages are generally darker in var. *janssoni.*

Distribution. Relatively common, found throughout Denmark and Fennoscandia. - Widespread throughout most of the Palearctic region; also in North America (introduced). - Var. *janssoni* is restricted to a limited area round the Baltic Sea (Sweden: Öl., Gtl. - Finland: Al, Ab, St.).

Biology. In all kinds of decaying organic matter, obviously showing a distinct synanthropy, thus especially found in various debris around farm buildings, such as compost heaps and barn manure. It is a very active flyer, also frequently flying in daytime, even in the very early spring. March-October; full grown larvae are found at the end of March, newly emerged adults in April; pupae and newly emerged adults also found at the beginning of September. Thus, the species probably has two generations per year.

74. *Cercyon quisquilius* (Linnaeus, 1761)
   Fig. 245, pl. 3: 6.

Scarabaeus quisquilius Linnaeus, 1761, Fauna Suec. (2): 138.

2.0-2.6 mm. Similar to *unipunctatus,* of almost the same colour, except that the elytra are always uniformly pale yellowish, without darker markings (only sometimes slightly darker round scutellum). Body shape (Fig. 245) a little narrower and more parallel-sided than in *unipunctatus.* Punctuation of head and pronotum as in *unipunctatus* (thus, slightly coarser and denser than in *terminatus*). Posterior angles of pronotum a little more rounded than in *unipunctatus.* Elytral striae a little finer, strial punctures also fine laterally; 10th stria very fine and indistinct. Interstices more finely punctured than in *unipunctatus.* Raised middle portion of mesosternum very narrow (index length: width about 5). Metasternum without femoral lines. Appendages reddish, antennal club darker, maxillary palpi usually almost uniformly reddish.

Distribution. Rather common; entire Denmark and Sweden; in Finland north to LkW; in Norway north to NTi.' – Widespread throughout most of the Palearctic region; also widely distributed in North America (introduced).

Biology. As *terminatus.*

75. *Cercyon terminatus* (Marsham, 1802)
   Fig. 244.

Dermestes terminatus Marsham, 1802, Ent. Brit. 1: 70.

1.6-2.3 mm. Piceous to black; elytra yellowish red to yellowish brown, normally with a large triangular blackish basal band, which is often fused basally with the narrowly dark anterior angles (as in Fig. 253); dark colour on dorsal portion of elytra sometimes extending further to the posterior, being separated medially by a narrow paler sutural stripe (except anteriorly); only seldom are the elytra almost entirely black. Head and pronotum finely and rather densely punctured. Anterior margin of clypeus truncate, not concave. Elytral striae rather fine, but distinct; inner striae distinct to base; 10th stria distinctly finer than 9th. Interstices usually with distinct microsculpture consisting of very fine irregular lines, finely and rather densely punctured anteriorly, the punctuation gradually becoming much finer posteriorly. Raised middle portion of mesosternum narrow (index length: width about 4). Metasternum without femoral lines. Appendages yellowish red, antennal club slightly darker. In elytral colour pattern often resembling *melanocephalus,* but easily recognized from it by the characters given in the key, the paler maxillary palpi and the smaller size.

Distribution. Widespread and rather common throughout most of Denmark and Fennoscandia. – Widely distributed in Europe, especially towards the north; eastwards to West Siberia; south to the Mediterranean, including North Africa (Algier); the Caucasus. Also introduced to North America where it is widespread (except in the northern parts).

Biology. Very euryoecious, in all kinds of decaying organic matter, mainly in dung (fresh cow and horse dung), decaying plant debris (e.g. compost heaps), etc. It is often abundant in various debris (e.g. barn manure) around farm buildings. The species is an active flyer. Adults are found in March-October, December.

76. *Cercyon pygmaeus* (Illiger, 1801)
Figs 227, 240.

Sphaeridium pygmaeum Illiger, 1801, Mag. Ins. 1: 40.

1.4-1.8 mm. Very similar to *terminatus,* of about the same colour, but black spot at anterior angle normally much more extensive; specimens with more distributed dark colouring are without distinct paler sutural stripe. Punctuation of head and pronotum normally much finer than in *terminatus,* elytral striae finer, the inner striae (particularly 2nd) very weak anteriorly, often obsolete; 10th stria much finer than 9th, often obsolete. Raised middle portion of mesosternum narrower than in *terminatus* (index length: width almost 5) (Fig. 227). Metasternum with conspicuous femoral lines reaching the anterior angles of metasternum. Appendages yellowish red, antennal club and maxillary palpi (at least terminal segment) normally darkened.

Distribution. Common and widely distributed throughout Denmark and Fennoscandia, apparently less frequent in the northernmost provinces. – Widespread throughout the Palearctic region and North America (introduced).

Biology. As *terminatus,* but normally much more abundant.

77. *Cercyon atricapillus* (Marsham, 1802)
Figs 228, 239.

Dermestes atricapillus Marsham, 1802, Ent. Brit. 1: 72 (sp. no. 31).
Dermestes nigriceps Marsham, 1802, Ent. Brit. 1: 72 (sp. no. 34).

1.5-2.0 mm. A rather wide and only moderately convex species. Reddish to pitchy brown; head black; pronotum dark brown with paler lateral margins or uniformly paler or reddish; elytra reddish or yellowish brown, each elytron often with a slightly darker, large spot just behind middle. Head and pronotum finely and densely punctured. Anterior margin of clypeus truncate, not concave. Lateral margins of pronotum straight anteriorly, in posterior third very strongly rounded (lateral view), more so than in our other species (Fig. 239). Elytral striae rather fine, with very fine punctures, which however become distinctly coarser laterally; 10th stria strongly reduced and often obsolete. Interstices flat anteriorly, becoming more convex posteriorly, somewhat shining, with very fine, posteriorly almost indistinct punctuation, and with very weak or indistinct microsculpture consisting of very fine irregular lines. Elytra with sparse, extremely fine and inconspicuous hairs (only visible at higher magnifications). Raised middle portion of mesosternum very narrow (index length: width about 7 or even more) (Fig. 228). Metasternum with conspicuous femoral lines reaching anterior angles of metasternum. Appendages yellowish red.

Distribution. A rather uncommon and local species; there are records from most parts of Denmark and Sweden; in Finland north to ObS; from Norway scattered records north to TRi. – A cosmopolitan species, known from most of the Palearctic region, Africa, South and East Asia, and North and South America (not in the more northern parts of North America).

Biology. Like *unipunctatus* the species is obviously to some extent synanthropic; it is usually found in barn manure, compost heaps and droppings of various mammals (mainly cow and horse dung) around farm buildings. It is an active flyer, frequently flying also at daytime; readily coming to light. January, May-October.

78. *Cercyon granarius* Erichson, 1837
Fig. 248.

Cercyon granarium Erichson, 1837, Käf. Mark Brand. 1: 221.

1.8-2.5 mm. Black; elytral apex and sometimes pronotal sides vaguely reddish. Head and pronotum finely and rather densely punctured. Anterior margin of clypeus truncate, not concave. Elytra short and wide, only about 1/7 longer than wide, greatest width well before middle, strongly convex. Striae very fine, mostly hardly impressed, with very fine, or laterally and near apex coarse punctures; 10th stria very conspicuous, not distinctly finer than 9th; the striae well defined to apex (Fig. 248). Interstices flat or (laterally) almost so, rather shining, especially laterally; with distinct, though very fine, reticulate microsculpture, and with very fine and rather sparse punctuation, which becomes extremely fine (or almost indistinct) laterally and posteriorly. Raised middle portion of mesosternum about twice as long as wide, with rather shining surface, separated posteriorly from anterior margin of raised metasternal portion by a narrow gap (as Fig. 234). Metasternum without femoral lines. Maxillary palpi brownish, often rather dark, antennae yellowish red with darker club, legs reddish to brownish.

Distribution. Very rare, only found at a few localities in Denmark (EJ, NWJ, F, LFM, SZ, NEZ); not in Sweden, Norway and Finland. – Central European species, generally rare or very rare; ranging from Denmark to the northern parts of South Europe; recorded eastwards to the Caucasus, westwards to Great Britain. – Records of the species from North America are due to misidentifications.

Biology. At the edge of fresh, probably mainly stagnant water, living among decomposing plant debris, such as leaf litter, perhaps mainly in well established, fairly eutrophic biotopes. Usually found singly or only few in number. Predominantly in spring.

Note. Lindroth (1960) mentions the species from a couple of Swedish provinces (Sk., Boh., Upl.). However, though many Swedish collections (private as well as museal) were examined, it was not possible to locate any *granarius;* all specimens standing as this species were misidentified. Thus, even though the species may occur in Sweden, it can not at the present time be considered as Swedish. A record from East Fennoscandia (Kangas, 1968a) apparently falls outside the main area of distribution, and needs confirmation.

79. *Cercyon tristis* (Illiger, 1801)
Figs 220, 234, 242, 247.

Sphaeridium triste Illiger, 1801. Mag. Ins. 1: 39.
Dermestes boletophagus Marsham, 1802, Ent. Brit. 1: 72.

1.9-2.5 mm. Very similar to *granarius* in colour and size, body shape a little less convex. Anterior margin of clypeus and punctuation of head and pronotum as in *granarius*. Elytra normally slightly or distinctly duller, with stronger reticulate microsculpture. The striae usually hardly impressed, strial punctures very fine anteriorly, becoming stronger (but not as strong as in *granarius)* laterally; hardly stronger apically. 10th stria very conspicuous, not distinctly finer than 9th; lateral striae well before apex becoming rather obsolete and somewhat irregular (Fig. 247). Interstices flat, very finely and rather sparsely punctured anteriorly, their punctuation becoming gradually finer (but not as fine as in *granarius*) laterally and posteriorly. Raised middle portion of mesosternum (Fig. 234) on the average a little narrower than in *granarius,* separated posteriorly from anterior margin of raised metasternal portion by a narrow gap. Metasternum without femoral lines.

Distribution. Common in entire Denmark; entire Sweden; South and Central Finland, north to ObS; from Norway scattered records north to STy; from North Fennoscandia the records are generally more sparse. – Widespread in North and Central Europe, ranging from France to Siberia. – Records of the species from North America are due to misidentifications.

Biology. At the edges of fresh, mainly stagnant, and usually rather eutrophic, waters, but fairly euryoecious, sometimes also at oligotrophic waters. It lives in wet mud, among wet moss, under decomposing plant debris (e.g. leaf litter), etc., both in open country and in woodland. Usually rather abundant. March-November.

80. *Cercyon convexiusculus* Stephens, 1829
Fig. 235.

Cercyon convexiusculum Stephens, 1829, Ill. Brit. Ent. Mandib. 2: 138.
Cercyon alni Vogt, 1968, Ent. Bl. Biol. Syst. Käfer 64: 186.

1.6-2.2 mm. In colour similar to the preceeding two species, but elytral apex is normally more pronouncedly reddish, and maxillary palpi are usually reddish (except for dark terminal segment). Punctuation of head and pronotum as in those species. Anterior margin of clypeus truncate, not concave. Elytral striae very fine, feebly im-

156

pressed, strial punctures very fine, becoming only slightly coarser laterally and apically; 10th stria distinct, a little finer than 9th; the striae also well defined apically. Elytra with fine and distinct, reticulate microsculpture, often rather dull, and with fine and somewhat sparse punctuation, which is normally distinctly stronger than in *granarius* and *tristis,* and does not become distinctly finer apically and laterally. Raised middle portion of mesosternum rather narrow (index length: width almost 3) (Fig. 235), its surface shining. Anterior margin of raised metasternal portion more prolonged anteriorly than in the two preceeding and the following species, contacting raised mesosternal portion. Metasternum without femoral lines.

Distribution. Rather common, found in most provinces of Denmark and Sweden; in Finland north to Ks; in Norway recorded from Ø and AK to Fn and Fø, but the records are very sparse. – Widespread in Europe, eastwards to the Caucasus and Siberia.

Biology. As *tristis,* and often found together with it.

81. *Cercyon sternalis* (Sharp, 1918)
Fig. 236.

Cerycon sternalis Sharp, 1918, Entomologist's mon. Mag. 54: 276.
Cercyon subsulcatus auctt.; *nec* Rey, 1885.

1.6-2.1 mm. Similar to *convexiusculus* in colour and size; maxillary palpi however normally almost uniformly pale reddish. Punctuation of head and pronotum as in the three preceeding species. Anterior margin of clypeus truncate, not concave. Elytral striae very fine, feebly impressed, normally distinctly deeper posteriorly; strial punctures very fine and shallow, laterally and near apex slightly coarser (apically often markedly so); 10th stria much finer than 9th, often obsolete. Interstices more dull than in the preceeding species, with very distinct reticulate microsculpture and with rather sparse and extremely fine punctuation, which is often very indistinct laterally and posteriorly. Raised middle portion of mesosternum at most about twice as long as wide (Fig. 236), with dull, markedly shagreened surface, posteriorly contacting anterior margin of raised metasternal portion. Metasternum without femoral lines.

Distribution. Relatively uncommon, only more frequent in southernmost Scandinavia; in East Denmark rather common, in Jutland only few specimens from SJ and EJ; local in South Sweden, north to Vrm.; in Finland only known from Al; not recorded from Norway. – South and Central European species, northwards to South Scandinavia; ranges from France to Yugoslavia and Hungary, perhaps eastwards to Siberia.

Biology. As *tristis,* and often found together with it.

Subgenus *Paracercyon* Seidlitz, 1888

Paracercyon Seidlitz, 1888, Fauna Balt. (2): 23.
Type species: *Hydrophilus analis* Paykull, 1798, by monotypy.

Pronotum not strongly convex, in lateral view forming a continuous curve with elytra. Raised middle portion of mesosternum rather wide posteriorly, closely received in a distinct v-shaped excision at anterior margin of metasternum (Fig. 237).

82. *Cercyon analis* (Paykull, 1798)
Figs 237, 243.

Hydrophilus analis Paykull, 1798, Fauna Suec. 1: 187.

1.7-2.5 mm. Black, dorsal surface shining; elytral apex, often also lateral margins (at least posteriorly) yellowish or reddish; sometimes lateral margins of pronotum are also paler. Head and pronotum finely and rather densely punctured, punctuation on head becoming very fine anteriorly. Anterior margin of clypeus truncate, not concave. Elytra widest in about anterior fourth to fifth, normally strongly narrowed posteriorly (Fig. 243) (on the average more so than in our other species); striae rather fine, anteriorly with rather fine punctures, the punctures becoming much coarser apically and laterally; 10th stria very distinct, at most slightly finer than 9th. Interstices flat or almost so, at most with indistinct microsculpture; punctuation anteriorly almost as on pronotum, becoming gradually much finer laterally and apically. Metasternum without femoral lines. Maxillary palpi and antennae yellowish red, antennal club darker; legs usually reddish brown, tarsi paler.

Distribution. Common to very common, found in the whole of Denmark and Fennoscandia. – Widely distributed throughout the Palearctic region; also widespread in North America (introduced).

Biology. In all kinds of decaying organic matter, especially decomposing plant debris (e.g. compost heaps); also on wet habitats near water, in wet moss, flood-debris, under leaf litter, etc.; apparently not in dung. It is an active flyer, often found in drift on the seashore, and readily coming to light. January-December; full grown larvae are found at the end of July, newly emerged adults in August.

Genus *Megasternum* Mulsant, 1844

Megasternum Mulsant. 1844, Hist. Nat. Col. Fr. Palp.: 186.
Type species: *Dermestes obscurus* Marsham, 1802 (Opinion 1178).

Small and broad, strongly convex species with strongly rounded sides. Dorsal surface glabrous. Clypeus truncate anteriorly, posteriorly not distinctly delimited from frons. Lateral margins of head, before eyes, rather abruptly narrowed, not hiding base of antennae; anterior margin of eyes not emarginate. Lateral margins of pronotum slightly inflexed, in lateral view almost evenly rounded (Fig. 256). Elytra with 10, usually distinct, regular series of punctures, and between the series with fine irregular punctuation; lateral margins inflexed so that elytra closely clasp the body. Epipleura very narrow, except near base (as Fig. 260). Ventral surface of body somewhat shining, without hydrofuge pubescence; mesosternum, metasternum and 1st abdominal sternite rather extensively punctured. Prosternum in middle strongly raised to form a broad, well defined flat plate, which is margined laterally, and markedly emarginate posteriorly to receive mesosternal apex (almost as Fig. 262). Middle portion of mesosternum raised to form a large well defined pentagonal plate, which is wider than long and broadly articulates with metasternum (as Fig. 261). Middle portion of metasternum only slightly raised and not sharply delimited. 1st abdominal sternite with a fine, sharp longitudinal ridge in middle. Antennae 9-segmented with compact club. Legs moderately long, tibiae flattened dorso-ventrally, with fine or very fine spines, protibiae on outer face with a pronounced excision in apical third (Fig. 257).

The genus comprises only 5 known species, represented in the Palearctic and Nearctic regions. Only one occurs in Europe.

83. *Megasternum obscurum* (Marsham, 1802)
    Figs 256-258; pl. 3: 9.

Dermestes obscurus Marsham, 1802, Ent. Brit. 1: 72.
Megasternum boletophagum auctt.; *nec* (Marsham, 1802).

1.7-2.0 mm. Reddish brown to almost black; lateral portions of pronotum variably paler; often also elytra partially paler, particularly apically. Dorsal surface rather shining, at most with extremely fine and rudimentary microsculpture. Head with dense and fine (on clypeus very fine) punctuation. Pronotum with more sparse and, especially on middle portion, very fine punctuation. Elytra with distinct series of fine or

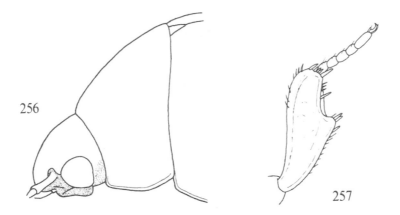

256

257

Figs 256, 257. *Megasternum obscurum* (Marsh.). – 256: head, pronotum and basal portion of elytra; 257: protibia and -tarsus.

159

very fine punctures; interstices flat, with rather sparse and extremely fine punctuation. Appendages yellowish red.

Distribution. Common or very common, found in entire Denmark and Fennoscandia. – Widespread in Europe and the Mediterranean area; also introduced to North America (restricted to a small area near Vancouver).

Biology. A very euryoecious species, living in all kinds of decaying organic matter, both on wet habitats and on more dry ground, in open country and in woodland; particularly in various kinds of decomposing plant debris, such as compost heaps, rotting grass, leaf litter or in moss, but also in rotting mushrooms, at carrion, at sap of deciduous trees and in dung. Sometimes – probably accidentally – found in nests of various mammals (mainly moles), or in company with ants *(Myrmica, Formica)*. Also taken in drift on the seashore. March-November.

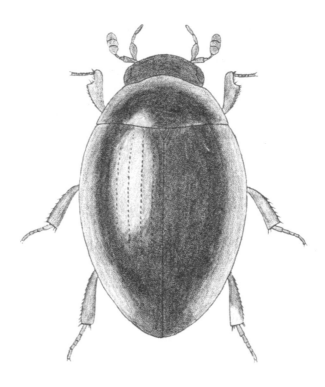

Fig. 258. *Megasternum obscurum* (Marsh.), length 1.7-2.0 mm. (After Victor Hansen).

160

Genus *Cryptopleurum* Mulsant, 1844

Cryptopleurum Mulsant, 1844, Hist. Nat. Col. Fr. Palp.: 186.
  Type species: *Sphaeridium minutum* Fabricius, 1775 (Opinion 1178).

In shape and size similar to *Megasternum*. Entire dorsal surface finely pubescent. Clypeus delimited from frons by a distinct transverse suture (which may be interrupted in middle); anterior margin of clypeus concave. Lateral margins of head, before eyes, rather abruptly narrowed, not hiding base of antennae; anterior margin of eyes not emarginate. Lateral margins of pronotum more inflexed than in *Megasternum,* in lateral view strongly angular in middle (Fig. 259). Elytra with 10 distinct punctured striae, 7th and 8th striae very close, often appearing as a single (only partially doubled) stria. Elytral interstices almost flat or distinctly convex, with irregular punctuation.

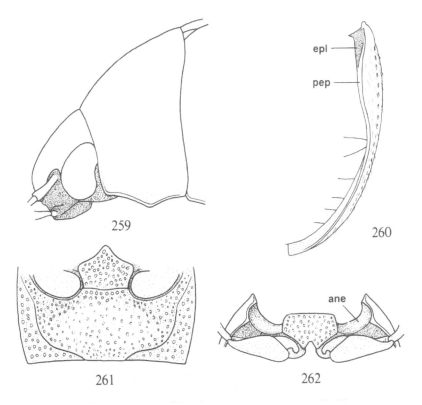

Fig. 259. Head, pronotum and basal portion of elytra of *Cryptopleurum minutum* (F.).
Figs 260-262. *Cryptopleurum crenatum* (Pz.). – 260: epipleural portion of elytra; 261: meso- and metasternum; 262: prosternum.

Lateral margins of elytra inflexed, and epipleura narrow (as in *Megasternum)* (Fig. 260). Ventral surface of body somewhat shining, without hydrofuge pubescence. Prosternum in middle raised to form a large, wide and flat, hexagonal plate, which is truncate anteriorly and markedly emarginate posteriorly to receive mesosternal apex (Fig. 262); prosternal plate not margined laterally. Meso- and metasternum almost as in *Megasternum,* except that raised mesosternal portion is larger and the (feebly) raised middle portion of metasternum is posteriorly and laterally sharply delimited by a curved, oblique femoral line (Fig. 261). Abdomen as in *Megasternum.* Appendages almost as in *Megasternum,* except for protibiae, which are without apical excision.

A genus of about 20 known species with a world wide distribution (missing only in the Australian region). 3 species occur in Europe.

Key to species of *Cryptopleurum*

1 Pronotum very finely punctured with distinct microsculpture consisting of fine longitudinal, somewhat irregular lines between the punctures. Ground colour usually reddish . 84. *subtile* Sharp
- Pronotum more strongly punctured, without distinct microsculpture (except on extreme lateral portions). Ground colour normally darker, piceous to black 2
2(1) Elytral striae somewhat depressed, interstices rather feebly convex. Punctuation of elytral interstices anteriorly as strong as pronotal punctuation . . . . 85. *minutum* (Fabricius)
- Elytral striae stronger depressed, interstices more convex, especially the lateral interstices, which are almost costate. Punctuation of elytral interstices much finer than pronotal punctuation . . . . 86. *crenatum* (Panzer)

84. *Cryptopleurum subtile* Sharp, 1884

Cryptopleurum subtile Sharp, 1884, Trans. R. ent. Soc. Lond.: 461.

1.5-2.2 mm Yellowish red to reddish brown with piceous to black head and sternites; normally also with darkened base and middle portion of pronotum; elytra sometimes partially darkened, but only very seldom as dark as in normally coloured specimens of *minutum.* Head extremely finely punctured; transverse furrow (separating clypeus and frons) complete, also in middle distinctly impressed. Pronotum very finely punctured (much finer than in *minutum).* Head and pronotum between the punctures with very distinct microsculpture consisting of fine, mostly longitudinal (but somewhat irregular) lines; the lines on clypeus however oblique or, at transverse furrow, transverse. Elytral striae almost as in *minutum,* except that 7th and 8th striae are distinctly separated and well defined (though still very close-set). Interstices distinctly, but rather feebly convex, the punctuation distinctly finer than in *minutum.* Appendages yellowish red, antennal club darker, tarsi and maxillary palpi often slightly paler.

♂: Last visible abdominal sternite at most slightly bulging apically.

♀: Last visible abdominal sternite with a distinct tubercle in middle near posterior margin.

Distribution. The species is a newcomer to the Scandinavian fauna; in recent decades it has rapidly dispersed, and is now comparatively common; it is known from a number of localities in Denmark, Sweden and Finland; in Norway only mentioned from AK and Ry. – Originally described from Japan (later discovered also on Taiwan); recently introduced to Europe (first recorded from Germany and Switzerland), where it is now widespread, at least in the northern and central regions; also widely distributed in North America (here possibly introduced directly from Japan).

Biology. In all kinds of decaying organic matter, particularly in compost heaps, barn manure and droppings of various mammals (e.g. horse dung); apparently somewhat thermophilic. The species is an active flyer, and frequently comes to light. April, July-October, perhaps mainly in autumn.

85. *Cryptopleurum minutum* (Fabricius, 1775)
   Figs 259, 263; pl. 3: 10.

Sphaeridium minutum Fabricius, 1775, Syst. Ent.: 68.
Sphaeridium atomarium sensu Olivier, 1790; *nec* Fabricius, 1775.

1.6-2.2 mm. Piceous to black, elytra usually becoming reddish apically, often also with a small, vaguely paler subhumeral spot; only seldom reddish with lateral and basal darkening, or uniformly reddish. Head and pronotum finely and densely punctured, head somewhat dull, often with obsolete microsculpture; the transverse suture rather widely interrupted in middle. Pronotum rather shining, without microsculpture, except on extreme lateral portions. Elytral striae somewhat impressed with rather fine punctures, that apically becomes obsolete or almost disappears; striae slightly deeper laterally and apically; 7th and 8th striae very close, appearing as a single (only in part doubled) stria. Interstices distinctly, but rather feebly convex, the punctuation anteriorly almost as on head and pronotum, becoming gradually much finer posteriorly. Abdominal sternites without secondary sexual characters. Appendages yellowish red, antennal club, maxillary palpi and usually femora darkened (maxillary palpi often brownish), tarsi rather pale.

Distribution. Common and widespread in the whole of Denmark and Fennoscandia. – Widely distributed throughout the Palearctic region; also introduced in North America (widespread, except in the southernmost warm areas).

Biology. In all kinds of decaying organic matter, usually very abundant, mainly in compost heaps, rotting grass, in dung and at carrion; also among various plant debris near water. An active flyer. April-October.

86. *Cryptopleurum crenatum* (Panzer, 1794)
   Figs 260-262.

Sphaeridium crenatum Panzer, 1794, Faun. Ins. Germ. (23): no. 3.

2.1-2.4 mm. Similar to *minutum* in colour and shape, but on the average larger. Head more shining and a little stronger punctured than in *minutum,* transverse furrow more strongly impressed, in middle often feebly so, but always complete. Pronotum shining without distinct microsculpture (except on inflexed portions); punctuation almost as on head (slightly stronger than in *minutum).* Elytral striae strongly impressed, with rather fine punctures, which become obsolete or almost disappear apically; the striae becoming deeper laterally and apically; 7th and 8th striae very close, appearing as a single (only in part doubled) stria. Interstices much more convex than in *minutum* (and *subtile*), especially the lateral insterstices, which are almost costate. Punctuation of interstices fine, anteriorly distinctly finer than on pronotum, laterally and apically becoming gradually very fine or even indistinct. Abdominal sternites without secondary sexual characters.

   Distribution. A rather rare or local species; found in most parts of Denmark; recorded from most provinces in Sweden and Finland (except the northern ones); in Norway only mentioned from Ø, On and TRi. – North and Central Europe, in general

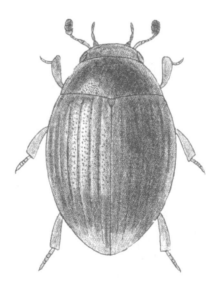

Fig. 263. *Cryptopleurum minutum* (F.), length 1.6-2.2 mm. (After Victor Hansen).

164

fairly local; southwards to southern France, Central Italy and the northern Balkans; ranging from France and Great Britain to the Caucasus.

Biology. In decaying organic matter, particularly droppings of various mammals, e.g. horse and cow dung, and in compost heaps, but also in wet habitats near water among plant debris, in moss, leaf litter, etc., perhaps mainly on open, sunny ground; obviously somewhat thermophilic. January, March-October.

SUBFAMILY HYDROBIINAE

Body contour evenly curved, not interrupted between pronotum and elytra. Head and pronotum without distinct impressions, surface rather evenly convex. Scutellum about as long as wide. Mesosternum in most forms distinctly raised medially, sometimes strongly so; mesosternal process not articulated to raised middle portion of metasternum. Abdomen with 5 visible sternites; seldom *(Laccobius)* the posterior margin of 5th visible sternite is broadly truncate or concave, so a 6th, more or less retractable sternite is exposed. Maxillary palpi as long as, or often distinctly longer than, the antennae, their 2nd segment not (or only a little) dilated. Antennae 7- to 9-segmented, with loose club. Tarsi 5-segmented with basal segment of meso- and metatarsi small, markedly shorter than 2nd segment; in *Cymbiodyta* it has disappeared so meso- and metatarsi are 4-segmented (basal segment thus long). Tibiae without fringes of long swimming hairs.

A large subfamily with a world wide distribution, comprising about 30 genera and a total of more than 700 known species. 8 genera occur in North Europe.

Key to genera of Hydrobiinae

1   Terminal segment of maxillary palpi distinctly longer than penultimate (Fig. 272); palpi only a little longer than the antennae (Tribe Hydrobiini) 2

–   Terminal segment of maxillary palpi shorter than, or at most as long as penultimate (Figs 305, 312, 313); palpi usually much longer than antennae (Tribe Helocharini) 6

2(1) Elytra without impressed sutural stria. Posterior margin of 5th visible abdominal sternite broadly truncate or concave, exposing a 6th, more or less retractable sternite (Fig. 282) .
. . . . *Laccobius* Erichson (p. 179)

–   Elytra with sharply impressed sutural stria, at least in posterior half. Abdomen with 5 visible sternites, posterior margin of 5th sternite simply rounded, not exposing a 6th retractable sternite 3

3(2) Elytra with distinct striae or series of punctures. Larger, 5.7-9.8 mm 4

– Elytra (besides sutural stria) without striae or series of punctures. Smaller, 2.3-3.3 mm 5

4(3) Mesosternum strongly raised in middle to form a high keel, the ventral margin of which is almost horizontal in lateral view (Fig. 277) .... *Limnoxenus* Motschulsky (p. 177)

– Mesosternum less strongly raised in middle, at most forming a postero-median process, the ventral margin of which is very oblique in lateral view (Figs 275, 276) .... *Hydrobius* Leach (p. 173)

5(3) 1st metatarsal segment much more than half as long as 2nd segment (Fig. 264). Antennae 8-segmented. Dorsal surface with faint metallic hue .... *Paracymus* Thomson (p. 167)

– 1st metatarsal segment less than half as long as 2nd segment. Antennae (in European species) 9-segmented. Dorsal surface not metallic .... *Anacaena* Thomson (p. 168)

6(1) Elytra without sutural stria .... *Helochares* Mulsant (p. 191)

– Elytra with sharply impressed sutural stria, at least in posterior half 7

7(6) All tarsi 5-segmented, with small basal segment (Fig. 265). Posterior margin of pronotum finely margined .. *Enochrus* Thomson (p. 195)

– Meso- and metatarsi 4-segmented (Fig. 266). Posterior margin of pronotum not margined .... *Cymbiodyta* Bedel (p. 207)

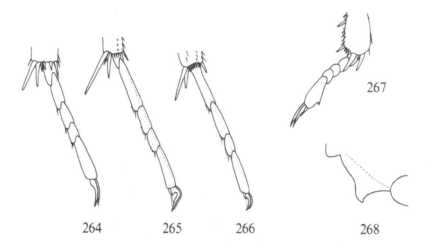

267

264   265   266   268

Figs 264-266. Metatarsi of Hydrobiinae. – 264. *Paracymus aeneus* (Germ.); 265: *Enochrus bicolor* (F.); 266: *Cymbiodyta marginella* (F.).
Figs 267, 268. *Paracymus aeneus* (Germ.). – 267: protarsus of male; 268: mesosternal process (lateral view).

166

Genus *Paracymus* Thomson, 1867

Paracymus Thomson, 1867, Skand. Col. 9: 119.
  Type species: *Hydrophilus aeneus* Germar, 1824, by monotypy.

Rather convex species. Head with feebly convex, hardly protruding eyes. Pronotum rather strongly narrowed anteriorly. Elytra with impressed sutural stria, reaching from apex anteriorly to well before middle, otherwise without distinct striae or series of punctures. Ventral surface of body somewhat dull, very finely shagreened and with dense hydrofuge pubescence. Prosternum raised in middle, tectiform. Mesosternum with a high sharp longitudinal ridge in middle, strongly dentiformly extended in later-

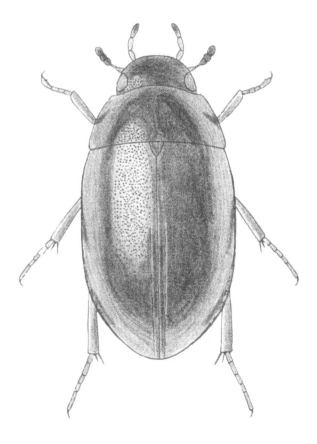

Fig. 269. *Paracymus aeneus* (Germ.), length 2.5-3.2 mm. (After Victor Hansen).

167

al view (Fig. 268). Middle portion of metasternum bluntly raised, at least posteriorly glabrous and shining. Abdomen with 5 visible sternites, posterior margin of last visible sternite entire, without emargination. Maxillary palpi about as long as antennae, terminal segment longer than penultimate. Antennae 8-segmented. Legs moderately long, all tarsi 5-segmented. Profemora (in at least basal half) and mesofemora (in at least basal third) finely and very densely punctured and pubescent, their distal portion smooth and shining, with very few and sparse punctures; metafemora entirely smooth and shining. Tibiae straight. Basal segment of meso- and metatarsi well over half the length of 2nd segment (Fig. 264). Protarsi slightly dilated in ♂, inner face of claw segment with a fine tooth in distal half (Fig. 267).

A large genus, comprising about 100 known species, represented in all major zoogeographical regions. One species occurs in Scandinavia.

87. *Paracymus aeneus* (Germar, 1824)
Figs 264, 267-269.

Hydrophilus aeneus Germar, 1824, Ins. spec. nov.: 96.

2.5-3.2 mm. Black, dorsal surface with distinct bronze or greenish hue. Entire dorsal surface with uniform, dense and rather coarse (on head a little finer) punctuation. Apex of mesosternal process slightly projecting backwards (lateral view) (Fig. 268). Appendages yellowish red, antennal club and femora (except at apex) darker; mesofemora densely and finely punctured and pubescent in about basal third.

Distribution. A rare species, only found in southernmost Scandinavia; in Denmark sparse records from SJ, NEJ, F, LFM, SZ and NEZ; in Sweden only in Sk.; in Norway only in AK; not in Finland. – Widely distributed along the coasts of the Mediterranean area (including North Africa) and western Europe to Scandinavia; also found at the Black Sea and the Caspian Sea, and on saline inland localities in Central Europe and South Russia.

Biology. A halobiontic species, occurring exclusively on salt marshes, at the edges of small shallow, well vegetated and often temporary pools above the high water mark. April-August, October.

Genus *Anacaena* Thomson, 1859

Anacaena Thomson, 1859, Skand. Col. 1: 18.
Type species: *Hydrophilus globulus* Paykull, 1798, by monotypy.

Similar to *Paracymus* in shape and size, though generally more convex. Eyes slightly convex, hardly protruding. Pronotum strongly narrowed anteriorly. Elytra with impressed sutural stria, reaching from apex anteriorly to well before middle, otherwise without striae or series of punctures. Ventral surface of body (at least partially) rather dull, very finely shagreened and with dense hydrofuge pubescence. Prosternum at most bluntly raised medially. Mesosternum not strongly raised in middle, but often with a small dentiform process or a small tubercle behind middle. Middle portion of metasternum feebly raised, at least partially glabrous and shining. Abdomen with 5 visible sternites, posterior margin of last visible sternite entire, without small apical excision. Maxillary palpi as long as or slightly shorter than antennae, terminal segment longer than penultimate. Antennae 9-segmented (in some exotic species 7- or 8-segmented). Legs moderately long, all tarsi 5-segmented. Femora very finely and densely punctured and pubescent, only near apex smooth and shining. Basal segment of meso- and metatarsi very short, well under half as long as 2nd segment. Protarsi at most feebly dilated in ♂.

A genus of about 50 known species, represented in the Palearctic, Nearctic, Neotropical and Oriental regions. 3 species occur in Fennoscandia and Denmark.

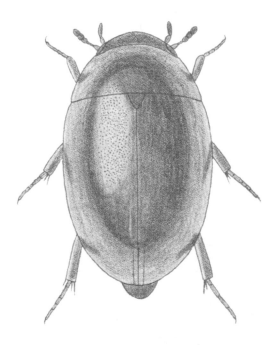

Fig. 270. *Anacaena limbata* (F.), length 2.3-3.0 mm. (After Victor Hansen).

Fig. 271. Mesosternum of *Anacaena lutescens* (Stph.) (ventro-lateral view).

Fig. 272. Maxillary palpus of *Anacaena*. – A: *lutescens* (Stph.); B: *globulus* (Payk.).

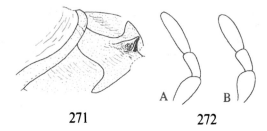

271   272

Key to species of *Anacaena*

1   Mesosternum at most with a small transverse tubercular bulge postero-medially. Terminal segment of maxillary palpi about 3 times as long as wide (Fig. 272B), normally yellowish red at least basally. Very broad and convex species .... 88. *globulus* (Paykull)
-       Mesosternum with a fine acute postero-median process (Fig. 271). Terminal segment of maxillary palpi about 4 times as long as wide (Fig. 272A), almost uniformly piceous to black. Narrower and less convex species 2
2(1) Head entirely black. Glabrous apical portion of metafemora larger, delimited from the pubescent portion by an oblique line (Fig. 273B) .... 89. *lutescens* (Stephens)
-       Head black with lateral margins narrowly reddish or yellowish in front of eyes. Glabrous apical portion of metafemora smaller, delimited from the pubescent portion by a less oblique line (Fig. 273C) .... 90. *limbata* (Fabricius)

88. *Anacaena globulus* (Paykull, 1798)
    Figs 272B, 273A.

Hydrophilus globulus Paykull, 1798, Fauna Suec. 1: 188.

2.7-3.3 mm. A short and wide, strongly convex species. Black; lateral margins of pronotum and elytra broadly paler; pale elytral margins often with series of small black spots. Dorsal surface shining, without metallic reflections. Entire dorsal surface with uniform, fine and (especially on head) dense punctuation. Mesosternum rather flat, with a small tubercular bulge postero-medially, that is sometimes very feeble. Maxillary palpi and antennae yellowish red, antennal club brownish; also terminal segment of maxillary palpi usually darker, but normally pale at least basally; the latter segment about 3 times as long as wide (Fig. 272B). Legs brownish red. Glabrous apical portion on ventral face of metafemora larger than in the two following species, sharply delimited from the pubescent portion by an oblique sinuate line (Fig. 273A).

Distribution. Very common in Denmark; widespread in Sweden north to Med., and in Norway north to NnV; in Finland only few specimens from Ab. – Widespread throughout the Palearctic region.

Biology. Predominantly in running waters, usually at the grassy edges of these, among submersed vegetation in shallow water; also in stagnant, perhaps especially acid waters, typically *Sphagnum*-pools. Occasionally found in drift on the seashore. January, March-November, most abundant in spring, but also frequent in late summer and autumn.

89. *Anacaena lutescens* (Stephens, 1829)
   Figs 271, 272A, 273B.

Hydrobius lutescens Stephens, 1829, Ill. Brit. Ent. Mandib. 2: 134.
Philydrus nitidus Heer, 1841, Fauna Col. Helv. 1: 485.
Anacaena limbata auctt.; *nec* (Fabricius, 1792).

2.4-3.1 mm. Body shape narrower and a little less convex than in *globulus*. Dorsal surface shining, without metallic reflections. Black, head without paler margins; pronotum and elytra reddish brown, usually rather dark. Pronotum with a very large, vaguely defined blackish median spot (often are only the lateral margins paler). Elytra with a narrowly darkened sutural stria, and with the punctures variably darker; sometimes are the elytra almost blackish, being paler only towards the margins (as in *globulus*). Punctuation of dorsal surface almost as in *globulus,* though somewhat variable (strongest in dark specimens). Mesosternum with a small acute postero-median process (Fig. 271). Maxillary palpi usually darker than in *globulus,* often reddish brown, and almost always with entirely dark (piceous to black) terminal segment, that is about 4 times as long as wide (Fig. 272A). Antennae yellowish red with brownish club. Legs reddish brown, femora (except apically) often darker. Ventral surface of metafemora with glabrous apical portion smaller than in *globulus,* sharply delimited

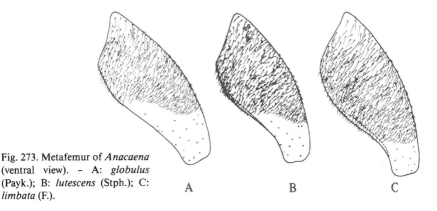

Fig. 273. Metafemur of *Anacaena* (ventral view). – A: *globulus* (Payk.); B: *lutescens* (Stph.); C: *limbata* (F.).

A   B   C

171

from the pubescent portion by an oblique, hardly sinuate line (Fig. 273B). Male protarsi slightly dilated, the claw segment inflated distally, with distinctly angular inner face, that bears a few stronger spines.

Distribution. A very common species, known from entire Denmark and Fennoscandia, except for the extreme north. – According to Berge Henegouwen (1986) it is widespread and common throughout Europe south to the Mediterranean (inclusive North Africa), ranging from West Europe to East Europe, probably further east. Also in North America (USA).

Biology. Very euryoecious, in almost all kinds of fresh water, predominantly in well vegetated pools, in moss, among vegetation etc., but also at the slower reaches of running water. It is an active flyer, often found in drift on the seashore. March-November; larvae are found in July, pupation in July-August, newly emerged adults in July and August. Males of this species are very rare, and all Fennoscandian and Danish specimens I have examined were females. Perhaps the species is chiefly parthenogenetic.

90. *Anacaena limbata* (Fabricius, 1792)
    Figs 270, 273C; pl. 3: 11.

Sphaeridium limbata Fabricius, 1792, Ent. Syst. 1 (1): 82.
Hydrobius ochraceus Stephens, 1829, Ill. Brit. Ent. Mandib. 2: 134.

2.2-2.8 mm. Body shape as in *lutescens*. Dorsal face usually paler. Head black, with lateral margins always narrowly reddish or yellowish in front of eyes. Pronotum and elytra yellowish to reddish brown. Pronotum with a rather small piceous or blackish median spot and an adjacent dark spot on each side of this; the three spots often mutually fused in anterior half (so pronotum is dark with only the margins and a basal spot on each side paler). Elytra with narrowly darkened sutural stria, and with the punctures variably darker; additionally a short row of blackish punctures is often present in middle near the sutural stria. Punctuation of dorsal surface on the average slightly finer than in the two preceeding species. Mesosternum as in *lutescens*. Colour of legs and antennae as in *lutescens;* maxillary palpi (except the dark terminal segment) rather pale, often yellowish, the terminal segment about 4 times as long as wide (as Fig. 272A). Ventral surface of metafemora with glabrous apical portion smaller than in the preceeding species and delimited from the pubescent portion by a less oblique line (Fig. 273C). Male protarsi only very feebly dilated, inner face of claw segment bearing a few stronger spines, but not angular.

Distribution. Rather common in southernmost Scandinavia; most parts of Denmark (though only sparse records from Jutland); in Sweden so far only known from Sk.; not in Norway and Finland (see note). – Widespread throughout the Palearctic region (no records from North Africa). Records from the more eastern parts of the Palearctic region may, at least in part, concern *lutescens,* and need to be confirmed.

Biology. In stagnant or sometimes slowly running fresh water, predominantly in well vegetated, eutrophic pools among vegetation, perhaps mainly on open ground. It seems to avoid more oligotrophic or acid waters. Occasionally it is found in brackish water. March-October. Contrary to *lutescens,* males of *limbata* are not markedly rarer than females.

Note. Most of the specimens standing as *A. limbata* in the Scandinavian collections have shown to belong to *A. lutescens,* which is a distinct species as pointed out by Berge Henegouwen (1986). On the basis of specimens from various parts of Europe (apart from Scandinavia) Berge Henegouwen has examined the West Palearctic distribution of the two species.

Genus *Hydrobius* Leach, 1815

Hydrobius Leach, 1815, *in* Brewst. Edinb. Enc. 9: 96.
 Type species: *Dytiscus fuscipes* Linnaeus, 1758, by monotypy.

Medium-sized, somewhat elongate, moderately convex species. Head with somewhat convex, but only feebly protruding eyes; clypeus with an irregular transverse row of coarse setiferous punctures on each side; frons with an oblique group of similar punctures on each side inside eyes. Pronotum strongly narrowed anteriorly, on each side with two, somewhat irregular oblique transverse, arcuate rows (one behind the other) of similar punctures. Elytra with 10 regular series of punctures, that (in the species treated here) stand in distinctly impressed striae; a short intercalary (sometimes obsolete) stria is present at base between 1st and 2nd striae; interstices almost flat, with irregular punctuation; 3rd, 5th, 7th and 9th interstices each with a row of sparse and coarse setiferous punctures, similar to those of head and pronotum; sometimes these setiferous elytral punctures are situated in 3rd, 5th, 7th and 9th striae, respectively, rather than in the interstices. Ventral surface of body very finely shagreened and with dense and fine punctuation and hydrofuge pubescence; more shining only on middle (feebly raised) portion of metasternum. Prosternum bluntly raised in middle, not distinctly ridged. Mesosternum posteriorly raised in middle , often forming a strong, acute process, ventral margin oblique in lateral view (Figs 275, 276). Metasternum with a short fine ridge anteriorly between mesocoxae, the ridge not continued posteriorly. Abdomen with 5 visible sternites, posterior margin of last visible sternite with a small emargination in middle (only very seldom, and not in the species treated here, without emargination). Maxillary palpi about as long as antennae, their terminal segment longer than penultimate. Antennae 9-segmented. Legs moderately long and slender, all tarsi 5-segmented. Femora, at least in basal half, finely and very densely punctured and pubescent, distally smooth and shining, and at most with very fine and sparse punctures. Tibiae straight. Dorsal face of meso- and metatarsi with a fringe of long fine swimming hairs, their basal segment minute, less than half as long as 2nd segment.

A small genus represented in the Palearctic, Nearctic, Afrotropical and Australian regions. It comprises only about 10 species, two of which occur in North Europe.

Key to species of *Hydrobius*

1   Mesosternum with a strong or rather strong acute dentiform process postero-medially (Fig. 275). Punctuation of elytral interstices normally fine or (subapically) very fine; strial punctures fine, often somewhat obsolete posteriorly (especially in the inner striae); the striae rather sharply impressed subapically . . . . 91. *fuscipes* (Linnaeus)

-   Mesosternum only bluntly raised postero-medially, without

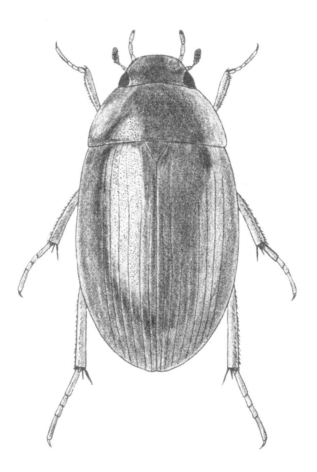

Fig. 274. *Hydrobius fuscipes* (L.), length 6.0-8.0 mm. (After Victor Hansen).

174

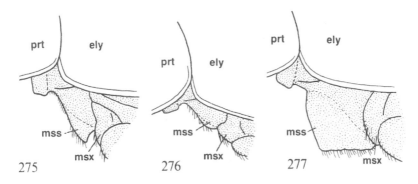

Figs 275-277. Mesosternal process in lateral view of Hydrobiinae. – 275: *Hydrobius fuscipes* (L.); 276: *H. arcticus* Kuw.; 277: *Limnoxenus niger* (Zschach).

strong acute process (Fig. 276). Punctuation of elytral interstices and striae stronger, the strial punctures also subapically
rather strong, but the striae less sharply impressed .... 92. *arcticus* Kuwert

91. *Hydrobius fuscipes* (Linnaeus, 1758)
   Figs 274, 275; pl. 4: 2.

Dytiscus fuscipes Linnaeus, 1758, Syst. Nat. (10) 1: 411.

Var. *subrotundus* Stephens, 1829:
Hydrobius subrotundus Stephens, 1829, Ill. Brit. Ent. Mandib. 2: 128.
Hydrobius picicrus Thomson, 1883, Annls Soc. ent. Fr. (6) 3: 131.
Hydrobius subrotundatus auctt.

Var. *rottenbergi* Gerhardt, 1872:
Hydrobius rottenbergi Gerhardt, 1872, Z. Ent. 3: 3.

6.0-8.0 mm. Piceous to black; lateral margins of pronotum and elytra often paler; abdominal sternites with reddish lateral spots. Dorsal surface shining, often with distinct
bronze, purplish or greenish hue. Head and pronotum densely and moderately finely
punctured, with some coarser setiferous punctures (cf. generic description). Elytral
striae fine, distinctly and (posteriorly) rather sharply impressed, anteriorly often very
feeble; strial punctures fine, often somewhat obsolete posteriorly (especially in the inner striae). Punctuation of elytral interstices a little finer than on head and pronotum,
becoming gradually finer posteriorly, very fine subapically. Alternate interstices with
rows of coarse setiferous punctures (cf. generic description). Mesosternum with a
strong or rather strong acute dentiform process postero-medially (Fig. 275). Appendages reddish; apex of maxillary palpi, antennal club and femora (except apically) darker.

175

Var. *subrotundus* Stph. is characterized by shorter and more convex body shape than in the typical form, and has darker legs, the tibiae being brownish to blackish, the femora being entirely dark; position of setiferous elytral punctures as in the typical form.

Var. *rottenbergi* Gerh. differs from the typical form and var. *subrotundus* by the position of the setiferous elytral punctures, that are situated in 3rd, 5th, 7th and 9th striae, rather than in the interstices; appendages reddish, as in the typical form.

Distribution. A very common species, known from entire Denmark and Fennoscandia. – Widely distributed throughout the Palearctic region; also in North America.

Biology. Very euryocious, mainly in stagnant water, but also at the slower reaches of streams, in both fresh and brackish water; in eutrophic water normally very abundant, but also rather frequent in more oligotrophic waters. It is usually found among vegetation on shallow places near the edge. Mainly in spring and autumn; larvae are found at the end of May, pupae and newly emerged adults at the beginning of July. The species is an active flyer, and is often found in drift on the seashore, and at light. – The typical form apparently prefers stagnant water, perhaps mainly neutral or basic fresh water; var. *subrotundus* is mainly found in more acid waters (e.g. *Sphagnum*-pools) or at the edges of running waters; var. *rottenbergi* seems to be confined to brackish water.

Note. The forms *subrotundus* and *rottenbergi* have now and then been given specific rank, e.g. by Lindberg (1943), who discussed the status of the three forms, based on Scandinavian material. Unfortunately, he did not include the following species *(arcticus)*, at least not under that name, and his conclusions should be considered inadequate. Of course, the forms may be good species, but no recent examinations satisfactorily confirm their specificity, wherefore I have preferred to consider them conspecific. As to the Fennoscandian distribution of the forms, *fuscipes* s. str. and *subrotundus* apparently occur throughout Denmark and Fennoscandia (the latter becoming predominant towards the north), while *rottenbergi* seems to be restricted to the coastal areas of South and Central Fennoscandia and Denmark. The Scandinavian distribution has not been examined in detail, just as an explanation of the Palearctic distribution of each form would require re-examination of most older records.

92. *Hydrobius arcticus* Kuwert, 1890
Fig. 276.

Hydrobius arcticus Kuwert, 1890, Bestimm. Tab. eur. Col. 19: 34.

5.7-6.5 mm. Very similar to *fuscipes*. Piceous to black, lateral margins of pronotum and elytra often paler. Body rather short (as in *fuscipes* var. *subrotundus*). Punctuation of head and pronotum about the same as in *fuscipes*. Elytral striae fine, distinctly impressed, the strial punctures rather strong, posteriorly distinctly stronger than in (normal specimens of) *fuscipes,* also conspicuous anteriorly. Punctuation of elytral interstices on the average a little stronger than in *fuscipes*. The rows of setiferous elytral punctures always situated in the interstices. Mesosternum bluntly raised postero-medially, at most very obtusely ridged, without strong acute process (Fig. 276). Metasternum a little shorter than in *fuscipes*. Appendages reddish, apex of maxillary palpi, antennal club, and femora (except apically) darkened.

Distribution. Apparently a fairly local species, known only from North Fennoscandia; in Sweden Ly. Lpm., T. Lpm. and Nb.; in Norway Nsi, TRi, Fi, Fn and Fø; in Finland ObN, Ks, LkW and Le; also Lr in the USSR; not in Denmark. – The species is not recorded outside Fennoscandia and the Kola Peninsula.

Biology. A typical tundra species from the alpine region. It lives in stagnant fresh water, particularly in grassy, often temporary pools (e.g. at *Carex*-tufts or in moss), and in *Sphagnum*-bogs; in Finland also found in a small stony pool without vegetation (I. Rutanen, *in litt.*). April-September, most numerously found in the summer.

Genus *Limnoxenus* Motschulsky, 1853

Limnoxenus Motschulsky, 1853, Hydrocanth. Russ.: 10.
Type species: *Hydrophilus oblongus* Herbst, 1797 (= *Hydrophilus niger* Zschach, 1788), by monotypy.

Very similar to, and in most characters concordant with *Hydrobius*. Differs by the medially strongly raised prosternum, that forms a high and sharp median ridge throughout its length; the medially very strongly raised mesosternum, that forms a narrow and high median keel throughout its length, with its ventral margin being almost horizontal in lateral view (Fig. 277). Metasternum more raised medially than in *Hydrobius*, with a rather strong ridge anteriorly between mesocoxae, the ridge continuing posteriorly to well behind mesocoxae. Elytra with 10 distinct series of punctures, 1st series (sutural stria) impressed posteriorly, the others not impressed. Head, pronotum and elytra with coarser setiferous punctures arranged as in *Hydrobius* (the elytral punctures always situated in the interstices). Metafemora, in the here treated species, only at extreme base with dense and fine pubescence and punctuation.

A small genus comprising only 7 known species, represented in the Palearctic, the Nearctic, the Afrotropical and the Australian regions. Only one species occurs in Europe.

93. *Limnoxenus niger* (Zschach, 1788)
   Figs 277, 278.

Hydrophilus niger Zschach, 1788, Mus. Lesk. 1: 35.

8.0-9.8 mm. Black, dorsal surface rather shining, often with a faint dark bronze or greenish hue. Entire dorsal surface with uniform, dense and fine punctuation, the punctuation on pronotum usually slightly coarser than on head and elytra. Elytra with distinct but fine series of punctures. Maxillary palpi reddish brown, the apex usually darkened; antennae yellowish red with darker club; legs brownish to black, tarsi reddish.

Distribution. Very rare in Denmark, only found in LFM (Skejten near Fuglsang, Gåbense) and SZ (Knudshoved); not in the rest of Fennoscandia. – Central and South

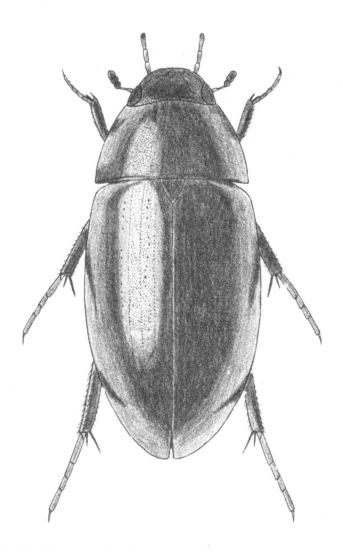

Fig. 278. *Limnoxenus niger* (Zschach), length 8.0-9.8 mm. (After Victor Hansen).

Europe, widespread in the Mediterranean (from Spain to Greece), eastwards to Siberia.

Biology. In stagnant, eutrophic water rich in vegetation, predominantly in fresh water; in Scandinavia exclusively found on warm, sunny sites near the coast. Thus, it is occasionally also found in brackish water, but the species is hardly halophilic. Rather, its restriction to coastal areas (in the northern parts of its range) is correlated with climatic factors. May-June, August-September.

Genus *Laccobius* Erichson, 1837

Laccobius Erichson, 1837, Käf. Mark Brand. 1: 202.
   Type species: *Chrysomela minuta* Linnaeus, 1758, by monotypy.

Generally wide and short, rather convex species with strongly rounded sides. Eyes feebly convex, at most slightly protruding. Pronotum normally strongly narrowed anteriorly, only seldom slightly posteriorly. Elytra without impressed sutural stria, but with numerous (about 20) close-set longitudinal series of punctures (Figs 279, 280); in some species *(gracilis)* the alternate series are markedly sparser and more finely punctured than the others; elytra only very seldom (and in no European species) with irregular punctuation. Ventral surface of body dull, very finely and densely shagreened and punctured and with dense hydrofuge pubescence; abdomen however, partially glabrous and shining (except at apex), in part coarsely punctured. Prosternum raised (often ridged) medially. Mesosternum strongly raised in middle to form a sharp ridge or keel, which is in lateral view angular or somewhat dentiform (Figs 286, 287). Middle portion of metasternum distinctly raised, shining postero-medially, glabrous and smooth. Abdomen (Fig. 282) with 6 visible sternites, posterior margin of 5th sternite truncate or more or less markedly concave, 6th sternite to some extent retractable. Maxillary palpi about as long as antennae, their terminal segment longer than penultimate. Antennae (Fig. 281) 8-segmented, the club widest at base. Legs moderately long, rather slender; femora glabrous and shining, only prefemora with large, densely and finely pubescent and punctured basal area. Metatibiae distinctly curved (except in a few exotic species). Posterior trochanters large. Tarsi 5-segmented, basal segment minute, less than half as long as 2nd segment. Meso- and metatarsi with a fringe of long swimming hairs on their dorsal face. 2nd and 3rd protarsal segments dilated in ♂ (Fig. 283).

A large genus with a world wide distribution (missing only in the Neotropical region), comprising about 150 known species; 9 of these occur in Scandinavia or the adjacent areas.

Key to species of *Laccobius*

1   Head yellowish, posteriorly (between eyes) and often also antero-medially darker. Pronotum yellowish with a pair of

rather small elongate dark spots in middle, the spots often
partially fused. Pronotum a little narrowed posteriorly. Ely-
tra about 1/3 longer than wide. (Fig. 290) . . . . 94. *decorus* (Gyllenhal)
-      Head piceous to black, at most with a pair of paler lateral
preocular spots. Pronotum not narrowed posteriorly, in
middle with a large dark (normally blackish) spot. Elytra
shorter, not 1/3 longer than wide (Figs 291, 292) 2
2(1) Small species, 1.8-2.4 mm. Serial punctures of elytra very
fine, rather inconspicuous, alternate series markedly more
sparsely punctured than the others . . . . *gracilis* Motschulsky
-      Larger, 2.5-4.0 mm. Serial punctures of elytra much more

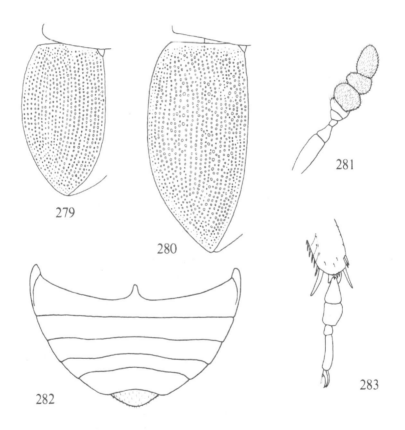

279

280

281

282

283

Figs 279, 280. Left elytron of *Laccobius*. - 279: *minutus* (L.); 280: *striatulus* (F.).
Figs 281, 283. *Laccobius sinuatus* Motsch. - 281: right antenna; 283: protarsus of male.
Fig. 282. Abdomen in ventral view of *Laccobius minutus* (L.).

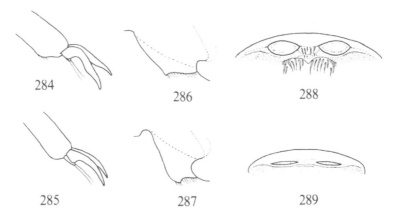

Figs 284, 285. Claws of *Laccobius*. – 284: *albipes* Kuw.; 285: *minutus* (L.).
Figs 286, 287. Mesosternal process in lateral view of *Laccobius*. – 286: *decorus* (Gyll.); 287: *minutus* (L.).
Figs 288, 289. Ventral surface of labrum of *Laccobius*-males. – 288: *sinuatus* Motsch.; 289: *bipunctatus* (F.).

conspicuous, the series uniform 3
3(2) Pronotum with fine, but distinct reticulate microsculpture 4
– Pronotum between the punctures smooth and shining, without distinct microsculpture 7
4(3) Claws strongly curved at base, distally almost straight, their inner face distinctly bisinuate (Fig. 284). Dark punctual spots of elytra generally small and well separated, elytra therefore seeming rather pale . . . . *albipes* Kuwert
– Claws evenly curved (Fig. 285). Dark punctual spots of elytra generally extensively confluent to form longitudinal lines, elytra therefore normally seeming much darker 5
5(4) Elytral series of punctures more or less irregular (as Fig. 280).
  ♂: Ventral face of labrum with a pair of narrowly transverse shining plates (Fig. 289). Length 3.0-3.5 mm . 99. *bipunctatus* (Fabricius)
– Elytral series of punctures regular, or almost so (Fig. 279).
  ♂: Labrum without such shining plates. On the average smaller, 2.6-3.2 mm 6
6(5) Pronotum with a very large transverse blackish spot in middle, that broadly touches anterior and posterior pronotal margins (Fig. 291). Elytra normally about 1/6-1/5 longer than wide. ♂: Aedeagus, Fig. 296 . . . . 95. *minutus* (Linnaeus)
– Dark middle spot of pronotum more narrowly touching pos-

terior margin (Fig. 292). Elytra on the average a little narrower, about 1/4 longer than wide. ♂: Aedeagus, Fig. 297
.... *cinereus* Motschulsky

7(3) Elytral series of punctures regular, or almost so (as Fig. 279). Elytra with a pair of very conspicuous pale subapical spots. Smaller, 2.5-3.1 mm .... 96. *biguttatus* Gerhardt

– Elytral series of punctures rather irregular (Fig. 280). Subapical elytral spots rather inconspicuous. Larger, 3.2-4.0 mm 8

8(7) Elytra pale yellowish, the punctual spots brownish and to some extent rather vague (especially anteriorly), hardly or only slightly confluent. Head black, without distinct paler preocular spots. ♂: Mesofemora (Fig. 295) posteriorly, near the trochanter, with a small patch of dense punctures and short, inconspicuous, yellow hairs. Aedeagus, Fig. 301 ....
.... 97. *sinuatus* Motschulsky

– Elytra normally seeming darker, the punctual spots black and normally very distinct, to a great extent confluent. Head (in Fennoscandian and Danish specimens) with a pair of distinct yellowish preocular spots. ♂: Mesofemora (Fig. 294) with a slightly larger patch of very dense punctures and longer, dense and rather conspicuous yellow hairs. Aedeagus, Fig. 300
.... 98. *striatulus* (Fabricius)

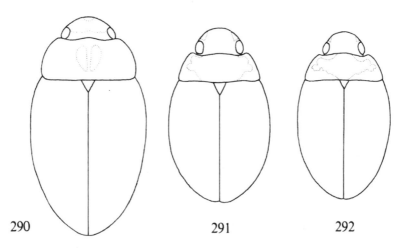

Figs 290-292. Outline of *Laccobius*. - 290: *decorus* (Gyll.); 291: *minutus* (L.); 292: *cinereus* Motsch.

Subgenus *Compsolaccobius* Ganglbauer, 1904

Compsolaccobius Ganglbauer, 1904, Käf. Mit. 4: 251.
Type species: *Hydrophilus decorus* Gyllenhal, 1827, by monotypy.

Elytral series of punctures uniform and regular, or almost so. Mesosternum strongly ridged medially, its ventral face in lateral view oblique, and only bluntly angular (Fig. 286). Pronotum (in the species treated here) widest a little before base, distinctly narrowed posteriorly, so body contour is slightly interrupted between pronotum and elytra.

94. ***Laccobius decorus*** (Gyllenhal, 1827)
  Figs 286, 290.

Hydrophilus decorus Gyllenhal, 1827, Ins. Suec. 4: 275.

3.0-3.5 mm. Yellow to reddish yellow, paler than our other species, meso- and

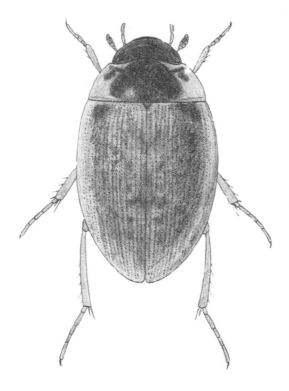

Fig. 293. *Laccobius sinuatus* Motsch., length 3.2-4.0 mm. (After Victor Hansen).

183

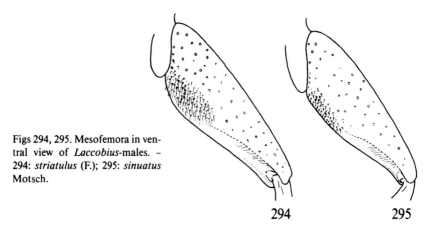

Figs 294, 295. Mesofemora in ventral view of *Laccobius*-males. – 294: *striatulus* (F.); 295: *sinuatus* Motsch.

294   295

metasternum (including coxae and episternae), and abdomen piceous to black. Head posteriorly (between eyes) and often also antero-medially darkened; pronotum with a pair of rather small and elcngate, close-set dark spots in middle; the spots are sometimes more or less fused, but a narrow pale median stripe is normally detectable, at least partially. Elytral punctures darkened (often blackish), except at lateral margins; dark punctual spots partially confluent longitudinally, particularly those of the alternate series. Dark portions of head and pronotum, and scutellum (sometimes also punctual elytral spots) often with a dark greenish tinge. Head and pronotum with fine or rather fine, not very dense and often somewhat uneven punctuation, between the punctures without or with only very indistinct microsculpture. Elytra narrower than in our other species, about 1/3 longer than wide. Appendages uniformly yellow or reddish yellow.

Distribution. Occurring round the Baltic Sea and the Gulf of Bothnia; in Sweden recorded from Sk., Sm., Öl., Gtl., Vg., Med., Ång. and Vb.; in Finland scattered records from the southern and western provinces north to ObN; not in Norway and Denmark. – Besides the European area of distribution (which includes Fennoscandia and the Baltic region of the USSR) the species is known from the Caucasus eastwards to Afghanistan and Central Asia.

Biology. A halobiontic species; in Scandinavia occurring exclusively along less exposed coasts, in brackish water with some vegetation and light, silty or clayey bottom; in the Caucasus and the Near East also found on saline inland localities. May-October, found very numerously in August.

Subgenus *Laccobius* Erichson, 1837, s. str.

Laccobius Erichson, 1837, Käf. Mark Brand. 1: 202.
   Type species: *Chrysomela minuta* Linnaeus, 1758, by monotypy.

184

Elytral series of punctures uniform and regular, or almost so (Fig. 279). Mesosternum very strongly ridged medially, forming a high keel, which is in lateral view sharply angular (often somewhat dentiform) (Fig. 287). Pronotum not narrowed posteriorly, body contour not interrupted between it and elytra.

Laccobius albipes Kuwert, 1890
Figs 284, 298.

Laccobius albipes Kuwert, 1890, Bestimm. Tab. eur. Col. 19: 82.

2.6-3.4 mm. Very similar to *minutus,* but with less extensive darkening. Head about the same colour as in *minutus,* dark with yellow preocular spots. Pronotum yellow with a large dark median spot, that narrowly touches anterior and posterior pronotal margins (almost as in Fig. 293). Elytra pale yellow, the punctures darkened (blackish), except at basal margin, a smaller area near scutellum, lateral margins and a small subapical spot on each elytron; the punctual spots small and generally well separated. Microsculpture of head and pronotum as in *minutus,* the punctuation normally distinctly finer. Appendages yellow. Claws (Fig. 284) strongly curved at base, distally almost straight, the inner face distinctly bisinuate (in all the other species treated here are the claws evenly curved)

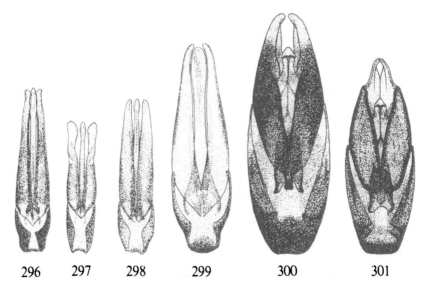

| 296 | 297 | 298 | 299 | 300 | 301 |

Figs 296-301. Male genitalia of *Laccobius* (dorsal view). - 296: *minutus* (L.), 297: *cinereus* Motsch.; 298: *albipes* Kuw.; 299: *bipunctatus* (F.); 300: *striatulus* (F.); 301: *sinuatus* Motsch.

♂: Mesofemora posteriorly, near the trochanter, with a small patch of denser punctures and fine yellow hairs (in ♀ simple). Aedeagus, Fig. 298.

Distribution. So far not recorded from Scandinavia. – A comparatively rare European species, ranging from France to the Caucasus, from the northern Mediterranean area to Central Europe, Poland, and perhaps as far north as North Germany (there are some old, unconfirmed records from Holstein); not in Great Britain. An occurrence in South Scandinavia is hardly likely, but can not be considered impossible.

Biology. Mainly at the edges of running waters, and obviously preferring sandy or silty banks of, perhaps mainly larger, rivers, but also found at the edges of temporary pools.

95. *Laccobius minutus* (Linnaeus, 1758)
Figs 279, 282, 285, 287, 291, 296; pl. 4: 5.

Chrysomela minuta Linnaeus, 1758, Syst. Nat. (10) 1: 372.

2.6-3.2 mm. Black; head on each side with a conspicuous yellowish preocular spot; pronotum with yellowish lateral margins, the dark middle spot very large, broadly touching anterior and posterior pronotal margins (Fig. 291); dark portions of head and pronotum often with a greenish or brownish tinge. Elytra yellowish, the punctures extensively darkened (blackish); punctual spots to great extent confluent to form longitudinal lines, except at basal margin, towards lateral margins, and in a vaguely demarcated subapical spot on each elytron. Head and pronotum with moderately dense, slightly uneven punctuation, which is rather fine on head (especially clypeus) and coarser on pronotum; between the punctures with fine, but distinct reticulate microsculpture, which sometimes extends posteriorly across anterior portion of elytra. Elytra short, with strongly rounded sides, only about 1/6-1/5 longer than wide (Fig. 291). Appendages yellowish red, apex of maxillary palpi, antennal club and profemora towards base, darker. Claws evenly curved (Fig. 285).
    ♂: Besides the dilated protarsi without secondary sexual characters. Aedeagus, fig. 296.

Distribution. Very common, found in entire Denmark and Fennoscandia. – Widely distributed throughout the Palearctic region.

Biology. In almost all kinds of fresh, occasionally also brackish waters, usually very abundant, especially in, or at the edges of shallow water with some, yet mostly rather sparse vegetation, and silty or clayey, rather light bottom; mainly on open ground. January, March-October. Larvae are found numerously at the end of June; pupae and newly emerged adults in July.

Laccobius cinereus Motschulsky, 1860
    Figs 292, 297.

186

Laccobius cinereus Motschulsky, 1860, Reis. Forsch. Amur-L. Schrenk 2 (2): 103.
Laccobius minutus var. *nanulus* Rottenberg, 1874, Berl. ent. Z. 18: 316.

2.6 mm. Very similar to *minutus,* but on the average a little smaller and narrower; elytra about 1/4 longer than wide (Fig. 292). Colour as in *minutus,* except that dark middle spot of pronotum only narrowly touches the posterior pronotal margin (Fig. 292). Punctuation of head, pronotum and elytra on the average slightly finer than in *minutus.* Microsculpture of head and pronotum as in *minutus,* or a little stronger, also in this species sometimes extending across anterior portion of elytra.

♂: Besides the dilated protarsi without secondary sexual characters. Aedeagus, Fig. 297.

Distribution. So far not recorded from Scandinavia. – Though mainly eastern palearctic, this species is found as far west as Archangelsk in NW. USSR, and in North Germany (Hannover and near Kiel), and may possibly occur in southern or eastern Fennoscandia.

Biology. Very little is known about the biology of this species. It seems to prefer stagnant waters, perhaps mainly in fresh (though also brackish) water; found in small pools or ponds, as well as in lakes.

96. *Laccobius biguttatus* Gerhardt, 1877

Laccobius biguttatus Gerhardt, 1877, Z. Ent. 6: 23.

2.5-3.1 mm. Similar to *minutus* in colour, size and shape. Dark middle spot of pronotum smaller, only rather narrowly touching anterior and posterior pronotal margins. Elytra towards base, especially round scutellum seemingly paler than in *minutus,* resulting from less extensive (or even indistinct) punctual darkening. Subapical pale spots of elytra very conspicuous, the punctures here only indistinctly darkened. Punctuation of head and pronotum on the average a little coarser than in *minutus,* between the punctures smooth and shining, without distinct microsculpture (on head sometimes with traces of reticulation). Appendages yellowish red, antennal club not darker.

Distribution. Rather common in most of Denmark, mainly near the coasts; in Sweden found in most provinces north to Upl. and Dlr., and in Nb.; in Finland recorded from most provinces north to Ok; not mentioned from Norway. – North and Central Europe, southwards to the northern Mediterranean, widespread in the USSR, eastwards to East Siberia (Yakutsk) and North China.

Biology. At the edges of fresh, neutral or basic, stagnant waters with some vegetation and silty or clayey, mostly rather light bottom; in Scandinavia obviously most abundant near the coasts, but – though occasionally found in slightly saline water – hardly halophilic, not living on the salt marshes. March-October, perhaps mainly in the spring.

Subgenus *Dimorpholaccobius* Zaitzev, 1938

Dimorpholaccobius Zaitzev, 1938 Trudý zool. Sekt., Baku 2: 120.
Type species: *Laccobius sulcatulus* Reitter, 1909, by monotypy.
Macrolaccobius Gentili, 1974, Memorie Mus. civ. Stor. nat. Verona 20: 550.
Type species: *Hydrophilus striatulus* Fabricius, 1801, by original designation.

Elytral series of punctures almost uniform, more or less irregular (Fig. 280). Mesosternum and body contour as in *Laccobius* s. str. Ventral face of labrum in ♂♂ (of the here treated species) with a pair of small transverse, lens-shaped or *(bipunctatus)* very narrow, shining plates (goggles, speculae) (Figs 288, 289), in ♀♀ without goggles.

97. *Laccobius sinuatus* Motschulsky, 1849
Figs 281, 283, 288, 293, 295, 301.

Laccobius sinuatus Motschulsky, 1849, Bull. Soc. Imp. Nat. Mosc. 22: 80.
Laccobius striatulus auctt.; *nec* (Fabricius, 1801).

3.2-4.0 mm. Head and ventral surface of body black; head without distinct paler preocular spots. Pronotum and elytra yellowish, normally very pale; pronotum with a large blackish median spot broadly touching anterior and posterior margins (though not as large as in *minutus)* (Fig. 293). Head and dark portions of pronotum often with a purplish, greenish of brownish tinge. Elytral punctures in general only slightly darkened (normally brownish), especially antero-medially and antero-laterally; the dark punctual spots hardly, or only slightly confluent. Elytra (besides punctual spots) often with some scattered larger vaguely darker spots; pale subapical elytral spots rather inconspicuous. Head and pronotum finely, not very densely punctured; head with distinct, but fine reticulate microsculpture, pronotum between the punctures smooth and shining, without microsculpture. Appendages yellowish red, profemora darker basally, antennal club at most slightly darker.

♂: Ventral face of labrum with a pair of transverse lens-shaped shining plates (Fig. 288). Mesofemora posteriorly, near the trochanters, with a small patch of dense punctures and short, rather inconspicuous yellow hairs (Fig. 295). Aedeagus, Fig. 301.

Distribution. Widespread, but rather local in Denmark, found in most provinces; in Sweden scattered records from a number of the southern provinces north to Upl. and Vstm.; in Finland only in the south (Al, N, and probably also St and Ta); not in Norway. – Widespread in Europe, from the South of England, Fennoscandia and the Baltic region of the USSR southwards to the Mediterranean (including Morocco, Algier and Tunisia), towards the east apparently not ranging so far to the south (Central Balkans); widespread in Central Europe, eastwards to the European USSR (perhaps ranging further east than *striatulus*). The species has been confused with *striatulus,* so its Palearctic distribution is still not known in details.

Biology. Predominantly at sparsely covered or bare edges of slowly running waters, especially in streams with clayey or silty, light bottom, in very shallow water or in wet mud at the edges. Often also in clay pits at groundwater pools and slow-flowing trickles. Sometimes found in drift on the seashore. April-July, September.

98. *Laccobius striatulus* (Fabricius, 1801)
Figs 280, 294, 300.

Hydrophilus striatulus Fabricius, 1801, Syst. Eleuth. 1: 254.

3.5-4.0 mm. Very similar to *sinuatus,* but usually recognized by the more pronounced darkening of the elytral punctures; these are normally blackish, and to a greater extent confluent. Further, the head has a pair of distinct yellowish preocular spots (the head may be entirely black, but apparently preocular spots are present in all Danish and Fennoscandian specimens). Appendages yellowish red, apex of maxillary palpi, antennal club, and profemora towards base, slightly darker.

♂: Labrum almost as in *sinuatus,* the shining plates a little less narrow. Mesofemora posteriorly, near the trochanters, with a small patch of very dense punctures and rather short yellow hairs, which are much more conspicuous than in *sinuatus* (Fig. 294). Aedeagus, Fig. 300.

Distribution. Comparatively local, found in most of eastern Denmark (all provinces except NWJ); in Sweden found sparsely north to Boh.; in Norway known only from a few southern provinces; not in Finland. – Widespread in Europe, ranging from the British Isles, South Scandinavia and the Baltic region of the USSR, southwards to the Mediterranean (with the exception of the Iberian Peninsula), from France to Asia Minor. Owing to confusion with *sinuatus* the Palearctic distribution is still not known in detail.

Biology. As *sinuatus* predominantly at the edges of slowly running waters with clayey or muddy bottom, but apparently preferring more well vegetated habitats, and more often found in shallow water among vegetation, rather than on bare clayey banks. Sometimes it, too, is found in clay pits at groundwater pools. Occasionally in drift on the seashore. April-June, August-October.

Note. Obviously all records from Finland (se Lindroth, 1960) concern *sinuatus.* I have examined specimens from Al and N, which are all *sinuatus,* and according to T. Clayhills *(in litt.)* all specimens in the collections of the Helsinki Museum are *sinuatus* (except for two *striatulus* from Kr in the USSR).

99. *Laccobius bipunctatus* (Fabricius, 1775)
Figs 289, 299.

Hydrophilus bipunctatus Fabricius, 1775, Syst. Ent.: 229.
Laccobius alutaceus Thomson, 1868, Skand. Col. 10: 311.

3.0-3.5 mm. Within the subgenus easily recognized by the fine, but distinct reticulate microsculpture of head and pronotum; otherwise – in shape, size and colour – often very similar to *minutus* in the preceeding subgenus, but recognized from it by the normally much more irregular elytral series of punctures. However, these series may be only somewhat irregular (though always less regular than in *minutus*), and slightly irregular in *minutus*. In such cases *bipunctatus* and *minutus* can usually be distinguished by distinctly finer punctuation on head, pronotum and elytra of the former, as well as by the primary and secondary ♂-characters.

♂: Ventral face of labrum with a pair of narrowly transverse shining plates (Fig. 289). Mesofemora simple. Aedeagus, Fig. 299.

Distribution. Rather common in Denmark; widespread in Sweden, but generally rather uncommon, particularly towards the north (northernmost record from Ång.); in Norway mentioned from the southernmost provinces and MRi; widespread in Finland north to Tb and Kb. – Widely distributed in Europe, ranging from Fennoscandia and the British Isles to the Mediterranean, including North Africa (Morocco), from France and Spain to the Balkans, Asia Minor and the European USSR.

Biology. Mainly at clayey or silty, often sparsely covered edges of stagnant, or sometimes slow-flowing fresh water; it often occurs in company with *minutus,* but is usually much less abundant.· January, March-October.

Subgenus *Microlaccobius* Gentili, 1974

Microlaccobius Gentili, 1974, Memorie Mus. civ. Stor. nat. Verona 20: 550.
  Type species: *Laccobius gracilis* Motschulsky, 1855, by original designation.

Differs from the preceeding subgenera in having the alternate elytral series of punctures markedly more sparsely punctured than the others. Mesosternum and body contour as in *Laccobius* s. str.

Laccobius gracilis Motschulsky, 1855

Laccobius gracilis Motschulsky, 1855, Etud. Ent. 4: 84.

1.8-2.4 mm. Easily recognized by the small size and the subgeneric characters. Ventral surface of body black, head dark, often blackish, with a pair of lateral rather small, pale preocular spots. Pronotum yellowish with large darker middle spot, that rather narrowly touches posterior pronotal margin; dark portions of head and pronotum often with a purplish or greenish tinge. Elytra yellowish, punctures (except towards elytral margins) darkened; the punctual spots hardly or only to a small extent confluent. Head and pronotum with fine and sparse punctuation, shining, without microsculpture. Elytra shining, the serial punctures very fine and rather inconspicuous. Appendages reddish yellow.

Distribution. So far not recorded from Scandinavia. – Widespread in the Mediterranean, scattered records from almost all of France, eastwards through Central Europe to Czechoslovakia and the Caspian Sea; other subspecies range further east (to North India) or further south (into the Afrotropical region). In recent time the species has declined, perhaps even disappeared from many parts of Central Europe; formerly it was found in North Germany (Hamburg), so the species may be found in southernmost Scandinavia, though this is perhaps not very likely.

Biology. Prefers the edges of slowly running water, but also at stagnant water, mainly pools situated near rivers or streams, occurring especially on habitats rich in filamentous algae. The species shows a pronounced preference for fairly warm waters, some exotic subspecies even being specialized in inhabiting hot springs.

Genus *Helochares* Mulsant, 1844

Helophilus Mulsant, 1844, Hist. Nat. Col. Fr. Palp.: 132 (nom. preocc.).
Helochares Mulsant, 1844, Hist. Nat. Col. Fr. Palp.: 197.
  Type species: *Dytiscus lividus* Forster, 1771, by subsequent designation (Thomson, 1859) (Opinion 710).

Eyes moderately convex, slightly protruding. Clypeus rather large, moderately strongly narrowed anteriorly, anterior margin broadly and markedly concave. Pronotum widest near base, somewhat narrowed anteriorly, but mostly not strongly so, sometimes also a little narrowed posteriorly; posterior margin not margined. Elytra moderately convex, in general widest a little behind middle; in the species treated here without striae; in other species with impressed sutural stria or even (as in many exotic species) with 10 punctured striae. Head, pronotum and elytra with coarser setiferous punctures, arranged almost as in *Hydrobius* (these punctures are, however, rather inconspicuous or hardly detectable in the species treated here). Ventral surface of body dull, very densely and very finely punctured, with dense hydrofuge pubescence. Prosternum feebly bulging in middle, but not ridged. Mesosternum with a posteromedian bulge (ridged only in some exotic species). Metasternum with raised middle portion, which is only vaguely delimited from rest of metasternum, not glabrous and shining. Abdomen with 5 visible sternites, posterior margin of last visible sternite with a small, often semicircular emargination in middle. Maxillary palpi (Fig. 305) long or very long, in the species treated here about twice as long as the antennae, their terminal segment a little shorter than penultimate. Antennae (Figs 303, 304) 9-segmented. Legs rather long and slender, femora with extremely fine and dense punctuation and pubescence, except near apex. All tarsi 5-segmented, basal segment of meso- and metatarsi very small, much less than half as long as 2nd segment. Claws in ♂ more curved and with much stronger basal tooth than in ♀ (Figs 306, 307).

A large genus with a world wide distribution, comprising about 150 species, mainly in the warmer climates. Only 2 species occur in Fennoscandia and Denmark.

A characteristic of *Helochares* is that the eggs are not placed in cocoons at the edges of the water, but are carried by the female in a bag on the ventral surface.

191

Key to species of *Helochares*

1　Elytra finely and densely punctured, with 2 (-3) rather clearly defined (often inconspicuous, but always detectable) rows of stronger setiferous punctures. Terminal segment of antennae at least twice as long as' wide (Fig. 303). ♂: Aedeagus, Figs 308, 310 . . . . 100. *punctatus* Sharp
–　Elytra a little more strongly and densely punctured, without (or almost without) distinct rows of stronger punctures. Terminal

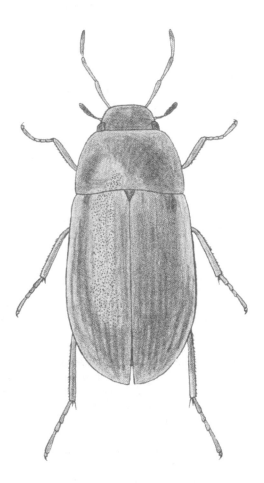

Fig. 302. *Helochares obscurus* (Müll.), length 4.7-5.9 mm. (After Victor Hansen).

segment of antennae shorter, not twice as long as wide (Fig.
304). ♂: Aedeagus, Figs 309, 311 .... 101. *obscurus* (Müller)

100. *Helochares punctatus* Sharp, 1869
Figs 303, 308, 310.

Helochares punctatus Sharp, 1869, Entomologist's mon. Mag. 5: 241.
Helochares lividus auctt.; *nec* (Forster, 1771).

4.9-6.1 mm. Dorsal surface reddish to yellowish brown, often rather dark. Ventral surface piceous to black. Head posteriorly and on middle portion of clypeus variably darkened, brownish to blackish. Labrum black, at most slightly reddish at anterior margin. Elytra sometimes with some incomplete longitudinal rows of small darker spots. Dorsal surface finely and, especially on pronotum densely punctured, rather shining. Head with some stronger, but rather inconspicuous punctures inside eyes and on antero-lateral portions of clypeus. Pronotum on each side with two transverse oblique rows of similar punctures, one behind the other. Elytra with similar, stronger punctures arranged in 2 (-3) rather clearly defined longitudinal rows, that may often be rather inconspicuous, but are always detectable. Appendages reddish, apex of maxillary palpi, antennal club and femora (except at apex) darker, brown to blackish. Terminal segment of antennae at least twice as long as wide (Fig. 303).

♂: Aedeagus (Figs 308, 310) with apex of parameres dilated inwards, forming a distinct obtuse angle (dorsal view). Basal margin of parameres sinuate (dorsal view). Membranous inner sac, in its natural position with several small dentiform bulges on each side, corresponding to a number of sclerotized spines, which become visible when the median lobe is extended.

Distribution. Only found sparsely in the South-west of Denmark (SJ, WJ); not in the rest of Fennoscandia. – A West European species; from Denmark to the British Isles, North and West France and North-west Spain, probably further south (to Morocco).

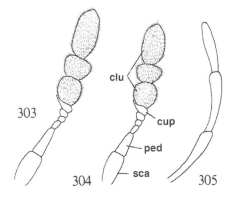

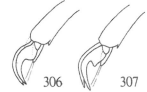

Figs 303, 304. Antennae of *Helochares*. – 303: *punctatus* Sharp; 304: *obscurus* (Müll.).
Figs 305-307. *Helochares obscurus* (Müll.). – 305: maxillary palpus; 306: protarsal claws of female; 307: protarsal claws of male.

Biology. More stenoecious than *obscurus,* predominantly (in Scandinavia exclusively) inhabiting the edges of acid and oligotrophic, fairly clear, stagnant waters, typically on moors. Mainly found in spring, but also in July-August.

101. *Helochares obscurus* (Müller, 1776)
Figs 302, 304-307, 309, 311; pl. 4: 1.

Hydrophilus obscurus Müller, 1776, Zool. Dan. Prodrom.: 69.
Helochares lividus auctt.; *nec* (Forster, 1771).
Helochares griseus auctt.; *nec* (Fabricius, 1787).

4.7-5.9 mm. Very similar to *punctatus,* differing in only a few external characters. Head a little paler than in *punctatus,* clypeus hardly darkened in middle. Labrum occasionally paler anteriorly. Elytra slightly more dull than in *punctatus,* a little coarser and more densely punctured, without (or with very indistinct) rows of stronger punctures. Setiferous punctures of head and pronotum detectable, but less distinct than in *punctatus.* Terminal segment of antennae shorter than in *punctatus,* not twice as long as wide (Fig. 304).

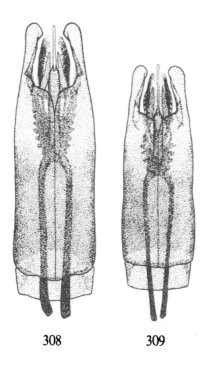

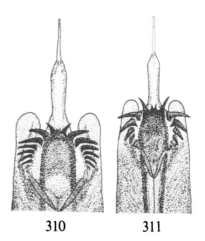

310   311

Figs 308, 309. Aedeagus of *Helochares,* with median lobe at rest within parameral tube. – 308: *punctatus* Sharp; 309: *obscurus* (Müll.). Figs 310, 311. Distal part of aedeagus of *Helochares,* with median lobe fully extended. – 310: *punctatus* Sharp; 311: *obscurus* (Müll.).

308   309

♂: Aedeagus (Figs 309, 311) smaller than in *punctatus,* inner face of parameres more rounded near apex, not distinctly angular (dorsal view). Basal margin of parameres evenly rounded, or straight, not sinuate (dorsal view). Membranous inner sac, in its natural position, with relatively larger, and fewer, dentiform bulges, corresponding to fewer sclerotized spines.

Distribution. Widespread and fairly common in Denmark and Central and South Fennoscandia; entire Denmark; Sweden north to Hls.; SE. Norway north to HEs; Finland north to Om and Kb. – A widely distributed species, ranging from Fennoscandia, SE. England and NE. France to North Italy, the Balkans and Asia Minor, eastwards to the Caucasus and Siberia.

Biology. Predominantly in stagnant, more or less eutrophic, often abundantly vegetated fresh water, mostly in neutral or basic water, but sometimes also in more acid waters, e.g. *Sphagnum*-pools; only very occasionally, however, on moors. March-October, mainly in spring; egg-carrying females are found mainly in May, small larvae at the end of May, pupae in July.

Genus *Enochrus* Thomson, 1859

Philydrus Solier, 1834, Annls Soc. ent. Fr. 3: 315 (nom. preocc.).
Philhydrus auctt.
Enochrus Thomson, 1859, Skand. Col. 1: 18.
   Type species: *Hydrophilus bicolor* Paykull (sensu Thomson, 1859 [nec *Hydrophilus bicolor* Fabricius, 1792] = *Hydrophilus melanocephalus* Olivier, 1792), by monotypy.

Similar to *Helochares,* but generally a little more convex. Pronotum rather strongly narrowed anteriorly, the posterior margin finely margined. Elytra with impressed sutural stria reaching from apex anteriorly to well before middle, but (in the species treated here) otherwise without distinct striae. Mesosternum (except anteriorly) strongly ridged in middle and narrowly raised to form an acute dentiform process. Metasternum on raised middle portion with a small postero-medial glabrous field. Abdomen with 5 visible sternites, posterior margin of last visible sternite entire, or with a small median emargination. Maxillary palpi quite characteristic (Figs 312, 313), much longer than antennae; their 2nd segment very long, with convex anterior (inner) face and concave (or almost straight) posterior face (in all the other genera of the subfamily oppositely curved, or straight and short); terminal segment shorter than, or as long as, penultimate, bending outwards (in the other genera bending inwards). Antennae 9-segmented. Legs almost as in *Helochares,* all tarsi 5-segmented, basal segment of meso- and metatarsi as in *Helochares.* Claws in ♂ at least a little more curved than in ♀, often with stronger basal tooth.
   A large genus with a world wide distribution, comprising about 200 known species; 10 of these occur in North Europe.

Key to species of *Enochrus*

1   Terminal segment of maxillary palpi as long as penultimate
(Fig. 312) .... 102. *melanocephalus* (Olivier)
–   Terminal segment of maxillary palpi distinctly shorter than
penultimate (Fig. 313) 2
2(1)   Larger, 4.7-6.8 mm. Posterior margin of last visible abdominal sternite entire 3
–   Smaller, 3.0-4.0 mm. Posterior margin of last visible abdominal sternite with a small semicircular emargination (Fig. 316) 8
3(2)   Head pale yellowish, frons only darkened in part. Maxillary
palpi uniformly reddish yellow .... 107. *bicolor* (Fabricius)
–   Head more extensively darkened, at least frons entirely black 4
4(3)   2nd segment of maxillary palpi pronouncedly and extensively
darkened in middle. Pronotum at most vaguely darkened in
middle. Large species, 5.7-6.8 mm .... 108. *testaceus* (Fabricius)
–   2nd segment of maxillary palpi uniformly pale, or if darkened
(very rare specimens of *fuscipennis),* pronotum with large
black median spot. Normally distinctly smaller species, 4.7-5.8 mm 5
5(4)   Elytra without any trace of rows of coarser punctures. Maxillary palpi uniformly pale (or almost so). Pronotum reddish to
yellowish brown, in middle with a very large, vaguely demarcated blackish spot. ♂: Claws rather strongly curved in
middle (Fig. 315), basal tooth of protarsal claws not markedly
larger than of the other claws .... 103. *ochropterus* (Marsham)

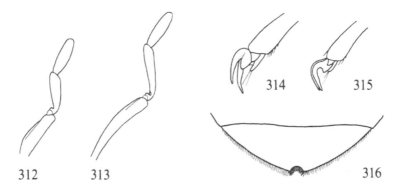

Figs 312, 313. Maxillary palpus of *Enochrus.* – 312: *melanocephalus* (Oliv.); 313: *ochropterus*
(Marsh.).
Figs 314, 315. Protarsal claws of *Enochrus*-males. – 314: *bicolor* (F.); 315: *ochropterus* (Marsh.).
Fig. 316. Last abdominal sternite of *Enochrus affinis* (Thbg.).

196

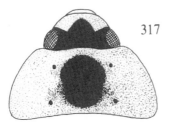 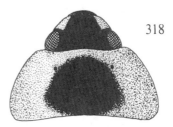

Figs 317, 318. Head and pronotum of *Enochrus*-males. - 317: *quadripunctatus* (Hbst.); 318: *fuscipennis* (Thoms.).

- Elytra with 2-3 distinct (often rather inconspicuous, but always detectable) rows of coarser setiferous punctures. Apex of maxillary palpi markedly infuscate, or if uniformly pale *(halophilus)*, pronotum in middle much less darkened, brownish. ♂: Claws strongly curved in middle, basal tooth of protarsal claws greatly enlarged (Fig. 314), much larger than of the other claws 6

6(5) Maxillary palpi uniformly pale. Pronotum rather vaguely darkened in middle, brownish (not black). ♂: Labrum and clypeus yellowish red, the latter with a black posteromedian triangular spot (as Fig. 317)...................... 106. *halophilus* (Bedel)

- Apex of maxillary palpi markedly infuscate. Pronotum darker in middle, with a variable blackish spot. ♂: Clypeus and labrum as in *halophilus*, or darker 7

7(6) Pronotum with 4 small black punctual spots arranged in a square, and between these with a larger black median spot, which does not normally fill up the field between these (Fig. 317). ♂: Labrum and clypeus yellowish red, the latter with a black postero-median triangular spot (Fig. 317) 105. *quadripunctaus* (Herbst)

- Black median spot of pronotum normally markedly larger, filling up completely the field between the 4 outer punctual spots (Fig. 318), sometimes even very large, leaving only lateral margins of pronotum paler. ♂: Labrum and clypeus black, the latter often with a pair of well marked yellowish preocular spots (Fig. 318); labrum only very seldom partially paler .... 104. *fuscipennis* (Thomson)

8(2) Terminal segment of maxillary palpi yellowish red, normally only slightly darker in middle (and not much darker than penultimate segment), always pale at apex. Head with a pair of conspicuous, large and well defined yellowish preocular

197

spots (Fig. 323). ♂: Median lobe of aedeagus wide (Fig. 320)
.... 110. *coarctatus* (Gredler)
- Terminal segment of maxillary palpi markedly darker than
penultimate, normally almost uniformly brown to blackish,
or at least dark distally. Head entirely black, or with distinct,
normally a little smaller preocular spots. ♂: Median lobe of
aedeagus narrow (Figs 321, 322)...................................... 9
9(8) Head entirely black, or almost so, only very seldom with distinct preocular spots. ♂: Aedeagus with apex of parameres
sharply pointed, and distinctly bending outwards (Fig. 322)
.... 109. *affinis* (Thunberg)
- Head with a pair of well marked yellowish preocular spots. ♂:
Aedeagus with apex of parameres more bluntly rounded, not
bending outwards (Fig. 321) .... *isotae* Hebauer

Subgenus *Enochrus* Thomson, 1859, s. str.

Enochrus Thomson, 1859, Skand. Col. 1: 18.
   Type species: *Hydrophilus bicolor* Paykull (sensu Thomson, 1859 [nec *Hydrophi-lus bicolor* Fabricius, 1792] = *Hydrophilus melanocephalus* Olivier, 1792), by monotypy.

Maxillary palpi rather short, terminal segment as long as penultimate (Fig. 312), posterior face of 2nd segment almost straight. Head, pronotum and elytra without distinct coarser setiferous punctures. Posterior margin of last visible abdominal ster-nite entire, without emargination. Claws in ♂ slightly stronger and more angularly curved than in ♀, and with stronger basal tooth.

102. *Enochrus melanocephalus* (Olivier, 1792)
   Fig. 312.

Hydrophilus melanocephalus Olivier, 1792, Enc. méth. 7: 127.

4.2-5.0 mm. Head black, on each side with a large yellow preocular spot, labrum black. Pronotum and elytra uniformly reddish yellow; pronotum with 4 small punctu-al black spots arranged in a square. Ventral surface of body black. Dorsal surface shin-ing, rather finely and densely, on elytra sligthly more sparsely punctured. Elytra (be-sides sutural stria) with rudiments of striae posteriorly (or rather, unimpressed rows of slightly stronger punctures), that must not be confused with setiferous punctures (present in subgenus *Lumetus*). Maxillary palpi only slightly longer than half the width of the head, reddish yellow, apex of terminal segment blackish. Antennae red-dish yellow with darker club. Legs dark brown to blackish, tarsi often a little paler.

   Distribution. A relatively rare or local species, found in most parts of Denmark (mainly near the coasts) and southern Sweden north to Upl. and Vstm.; southern Finland (Al, Ab, N, Ka, Oa); in Norway only mentioned from Ø, AK and VE. – Europe; from Fennoscandia to the Mediterranean (including North Africa), from France and the British Isles to Asia Minor.

Biology. At the edges of stagnant, neutral or basic, predominantly fresh and usually shallow, well vegetated waters with clayey bottom, in Scandinavia particularly near the coasts, typically in company with *Laccobius biguttatus;* as in this species, *E. melanocephalus* is hardly halophilic, rather to some extent thermophilic. March-May, July-October.

Subgenus *Lumetus* Zaitzev, 1908

Philydrus Solier, 1834, Annls Soc. ent. Fr. 3: 315 (nom. preocc.).
Philhydrus auctt.
Lumetus Zaitzev, 1908, Trudý russk. ént. Obshch. 38: 385.
    Type species: *Hydrophilus bicolor* Fabricius, 1792, by subsequent designation (d'Orchymont, 1939).

(The subgeneric name *Pseudenochrus* Lomnicki (1911: 266) used for *Enochrus ochropterus* (Marsham) is based merely on slight differences in elytral punctuation, and is here considered superfluous).

Maxillary palpi longer than in *Enochrus* s. str., terminal segment markedly shorter than penultimate (Fig. 313). Setiferous punctures of head and pronotum often rather inconspicuous (especially in *ochropterus*), but always detectable. Elytra (except in *ochropterus*) with a few distinct, though often rather inconspicuous rows of slightly coarser setiferous punctures. Posterior margin of last visible abdominal sternite entire, without emargination. Claws in ♂ distinctly stronger and more angularly curved than in ♀, with stronger basal tooth, which is in protarsi often greatly enlarged (Figs 314, 315).

103. *Enochrus ochropterus* (Marsham, 1802)
    Figs 313, 315.

Hydrophilus ochropterus Marsham, 1802, Ent. Brit. 1: 409.
Hydrobius frontalis Erichson, 1837, Käf. Mark Brand. 1: 210.

4.7-5.5 mm. Head in ♀ entirely black, or often with a pair of yellowish preocular spots, in ♂ with black frons and black posteromedian spot on the otherwise reddish clypeus; labrum black (♀) or reddish (♂). Pronotum reddish to yellowish brown, in middle with a very large, rather vaguely demarcated blackish spot. Elytra reddish to yellowish brown. Ventral surface of body black. Dorsal surface shining, rather finely and densely punctured, and except for setiferous punctures of pronotum (which are rather inconspicuous) without distinct coarser punctures. Appendages yellowish red;

apex of maxillary palpi at most indistinctly darkened; antennal club dark; femora (except at apex) at least partially darker.

♂: Claws rather strongly angularly curved in middle, with rather strong basal tooth (Fig. 315); in ♀ more evenly curved with smaller basal tooth. Colour of head sexually dimorphic (cf. above).

Distribution. Rather common, widespread throughout Denmark and Fennoscandia, yet from Norway only more scattered records. – North and Central Europe, south to Central France, North Italy and the northern Balkans, ranging from the British Isles to Siberia.

Biology. A tyrphophilic species, inhabiting stagnant fresh water, normally in more or less acid, often somewhat shaded waters in woodland, typically in *Sphagnum*-pools, often occurring in company with *affinis*. March-October.

104. *Enochrus fuscipennis* (Thomson, 1884)
Fig. 318.

Philhydrus fuscipennis Thomson, 1884, Opusc. Ent. 10: 1031.
Philydrus sahlbergi Fauvel, 1887, Revue Ent. 6: 86.
Enochrus quadripunctatus auctt.; *nec* (Herbst, 1797).

4.7-5.5 mm. A rather variably coloured species. Head and ventral surface of body black; head usually with a pair of distinct yellowish preocular spots, that may be very narrow or even missing (independent of the sex). Labrum black, only very seldom (in some specimens of var. *sahlbergi*, cf. below) partially paler. Pronotum yellowish brown to reddish brown, with a large variable blackish median spot (Fig. 318), that is normally somewhat trapezoid (narrowed anteriorly), and almost always fills up the field between the 4 small punctual black spots (compare with the following species), sometimes even larger, leaving only the lateral margins of pronotum paler. Elytra yellowish brown to reddish brown, often with a small black humeral spot, sometimes with darker colour, and in extreme specimens black with only lateral margins reddish. The variation in the colour of head is independant of pronotal and elytral colouration. Dorsal surface shining, punctuation almost as in *ochropterus*, but on the average slightly sparser, and with 2-3 detectable, though often rather inconspicuous rows of coarser punctures on each elytron. Appendages yellowish red; apex of maxillary palpi markedly infuscate, occasionally with almost uniformly dark terminal segment and/or darkened 2nd segment; antennal club darker; legs yellowish red, seldom darker, femora (except at apex) at least partially darker. Specimens with blackish pronotal and elytral ground colour and entirely (or almost entirely) black head have been described as *sahlbergi* Fauv., but do not seem particularly distinctive (see note under *quadripunctatus*).

♂: Claws strongly angularly curved in middle, with strong basal tooth, especially on protarsi (as Fig. 314). Colour of head not sexually dimorphic.

Distribution. Rather common and widespread, found throughout most of Denmark and Sweden (except Lapland); Finland north to Om; in Norway scattered records north to NTi. – Widespread in Europe, from Fennoscandia to the northern part of the Mediterranean, from Spain and France to East Europe, probably further east. – Var. *sahlbergi* is restricted to Fennoscandia, occurring round the Baltic Sea and the Bay of Bothnia (Sweden: Gtl., Ög., Sdm., Upl., Med. – Finland: Al, Ab, N, Ka, Ta).

Biology. At the edges of stagnant, usually well vegetated fresh water, often on somewhat shaded ground and mainly in acid waters, but apparently more euryoecious than *ochropterus,* and perhaps preferring a little more eutrophic water; occasionally also found in slightly saline water. April-October.

Note. *E. fuscipennis* is often considered a form of *quadripunctatus* (see discussion under *quadripunctatus*).

105. *Enochrus quadripunctatus* (Herbst, 1797)
Fig. 317.

Hydrophilus quadripunctatus Herbst, 1797, Natursyst. all. bek. in-u. ausl. Ins. 7: 305.

4.7-5.8 mm. Very similar to *fuscipennis,* and apart from the colour of the dorsal surface, concordant with it in apparently all external (as well as aedeagal) characters. Head in ♀ black with a pair of distinct yellow preocular spots (as Fig. 318), that are usually rather large and often very narrowly fused medially at anterior margin of clypeus; in ♂ with black frons and a black, sharply defined postero-median (almost always triangular) spot on the otherwise yellowish red clypeus (Fig. 317); labrum black (♀) or yellowish red (♂). Pronotum yellowish red to brownish yellow, normally paler than in *fuscipennis.* Pronotum (Fig. 317) with 4 small punctual black spots arranged in a square, and a rather large black median spot, which is normally narrower and markedly smaller than in *fuscipennis,* not filling up the field between the 4 punctual spots. Elytra (as in *fuscipennis*) often with a small black humeral spot. Maxillary palpi reddish yellow, the apex markedly infuscate, often blackish; antennae yellow with darker club; legs yellowish red, femora (except at apex) at least partially darker.

♂: Claws as in *fuscipennis.* Colour of head sexually dimorphic, in ♂ different from *fuscipennis-*♂ (cf. above).

Distribution. Widespread and rather common, mainly along the coasts of South and Central Fennoscandia; found in most provinces of Denmark; in Sweden north to Nb.; widespread in Finland (where it is not restricted to the coast) north to ObN; from Norway only a few specimens from Ø. – Widely distributed in the Palearctic region; owing to the confusion with related (?) species (perhaps *fuscipennis* in particular, see note), the palearctic distribution is not known in detail.

Biology. In stagnant water, especially open, shallow pools with some vegetation and more or less clayey bottom, in Scandinavia predominantly near the coast and perhaps to some extent halophilic, but also found in fresh water. It is a very active flyer, often found in drift on the seashore. April-October.

Note. *E. quadripunctatus* and its relatives (i.e. *fuscipennis, sahlbergi* and *halophilus*) have been – and for that matter still are a source of much confusion. Normally, all four forms are easily recognized (by their colour), but occasionally individuals are found that look intermediate between two of the forms (*sahlbergi* x *fuscipennis* or *fuscipennis* x *quadripunctatus*). Because the separation is based sole on differences in colour, and the male genitalia do not seem to provide any reliable characteristics, opinions vary as to whether the forms should be given specific rank. Some authors consider them conspecific, some them to represent four distinct species. Others recognize two species (*fuscipennis* (= *sahlbergi*) and *quadripunctatus* (= *halophilus*)), and still others (e.g. Lohse, 1971) give *halophilus* specific rank but include *fuscipennis* (and *sahlbergi*) under *quadripunctatus*.

In attempt to clear up the situation, I have examined a large number of specimens from Denmark and Fennoscandia. In regard to the colour of the upper side, *fuscipennis* appears to be the most variable form, ranging from yellowish to dark brown; very dark specimens can be extremely difficult to recognize from *sahlbergi*, which probably is merely a melanistic variety of *fuscipennis*. As to the very light *fuscipennis* they may, on the other hand, be very similar to *quadripunctatus*, but are almost always easily distinguished by a larger black pronotal spot (see description). Specimens appearing intermediate between the two are so rare, that they do not form the basis of rejecting a possible specific rank of the two (I have seen only a couple of such specimens (females) from SW. Finland). Therefore, based on the difference in the colouration of the head in males (see the description), which is obviously a very reliable character, and on the ecological difference between *fuscipennis* and *quadripunctatus*, I have here preferred to treat them as good species. As to *halophilus*, which – contrary to both *quadripunctatus* and *fuscipennis* – is an exclusive inhabitant of brackish water, it may be considered a pale variety of *quadripunctatus*, but is in fact quite constant. I have not seen any intermediates between it and *quadripunctatus*, and therefore follow the interpretation of Lohse (1971), considering *halophilus* a distinct species. This view is further emphasized by the ecological and distributional differences.

106. *Enochrus halophilus* (Bedel, 1878)

Philydrus halophilus Bedel, 1878, Annls Soc. ent. Fr. (5) 8: 169.

5.0-5.8 mm. Very similar to the two preceeding species, especially *quadripunctatus*. Colour as in *quadripunctatus*, rather pale, and showing the same sexual dimorphism in colour of head. Pronotum, however, not black in middle, but with a large brownish, rather vaguely demarcated spot, that fills up the field between the 4 black punctual spots. Punctuation of dorsal surface on the average a little finer than in *quadripunctatus* and *fuscipennis*. Maxillary palpi uniformly reddish yellow, not infuscate at apex. Colour of antennae and legs as in *quadripunctatus*.

♂: As *quadripunctatus*.

Distribution. In Scandinavia only known from the coasts of southern Denmark (F,

202

LFM, SZ, NEZ). – Distributed along the coasts of West Europe; France and Great Britain, south to Morocco; also recorded from Cyprus, and probably more widespread in the Mediterranean.

Biology. A halobiontic species, found exclusively on salt marshes along the coasts, in shallow, sparsely vegetated, often temporary brackish pools above high tide line. April-August.

Note. *E. halophilus* is here considered a distinct species, though some authors have referred to it as a brackish water form of *quadripunctatus* (see discussion under that species). Earlier is was considered a subspecies of *bicolor*, but it is no doubt more closely related to *quadripunctatus*.

107. *Enochrus bicolor* (Fabricius, 1792)
    Figs 265, 314.

Hydrophilus bicolor Fabricius, 1792, Ent. Syst. 1 (1): 184.
Philhydrus maritimus Thomson, 1853, Öfvers. K. VetenskAkad. Förh. 10: 51.

5.0-6.5 mm. Dorsal surface uniformly pale, reddish to brownish yellow. Head paler than in our other species; frons only darkened in part; clypeus uniformly pale, or (in some ♀♀) slightly darker medially; labrum pale, in ♀ often darkened, occasionally entirely black. Pronotum with 4 small punctual black spots arranged in a square, in middle only weakly darkened. Ventral surface of body blackish. Dorsal surface on the average a little more finely punctured than in the preceeding species; rows of coarser setiferous punctures on elytra distinct. Maxillary palpi uniformly pale, reddish yellow; colour of antennae and legs as in *quadripunctatus*.

♂: Claws as in *fuscipennis* (protarsal claws, Fig. 314). Colour of head showing only some tendency to sexual dimorphism (cf. above).

Distribution. Widespread along the coasts of Central and South Scandinavia; common in Denmark, and in Sweden north to Upl.; in Finland north to Oa; southern Norway north to STy. – Widely distributed along the European coasts, from Fennoscandia and the Baltic region of the USSR, throughout the coasts of West Europe, southwards to the Mediterranean, where it is widespread, eastwards to Asia Minor; also found at saline inland localities in Central and South-eastern Europe and the USSR.

Biology. As *halophilus* in brackish water near the coasts, but usually far more abundant; also in slow-flowing water, e.g. drainage canals. It obviously tolerates lower salinities, and is sometimes, although very occasionally, found in fresh water, and then mainly at the edges of lakes. April-October; larvae and pupae are found in both July and in September, newly emerged adults in August.

108. *Enochrus testaceus* (Fabricius, 1801)
    Fig. 319.

Hydrophilus testaceus Fabricius, 1801, Syst. Eleuth. 1: 252.

5.7-6.8 mm. Very similar to *bicolor*, of similar rather pale colour, except for the head and maxillary palpi. Frons always entirely black; clypeus pale with black posteromedian spot, which is in ♂ rather small (or even missing), in ♀ normally much larger, and often reaching almost to anterior clypeal margin; labrum in ♂ reddish, in ♀ usually darkened (at least partially). Punctuation of dorsal surface as fine as in *bicolor*, but a little denser. Maxillary palpi reddish yellow, pseudobasal (2nd) segment pronouncedly and extensively darkened in middle, apex of terminal segment hardly so.

♂: Claws as in *fuscipennis*. Colour of head sexually dimorphic, though not quite constant (cf. above).

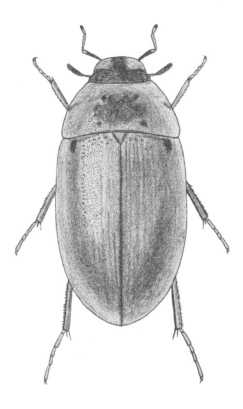

Fig. 319. *Enochrus testaceus* (F.), length 5.7-6.8 mm. (After Victor Hansen).

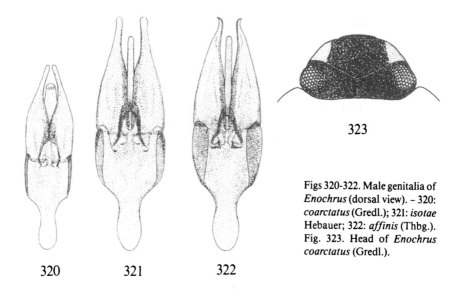

323

Figs 320-322. Male genitalia of
Enochrus (dorsal view). - 320:
coarctatus (Gredl.); 321: *isotae*
Hebauer; 322: *affinis* (Thbg.).
Fig. 323. Head of *Enochrus*
coarctatus (Gredl.).

320   **321**   **322**

Distribution. Widespread and fairly common in Denmark and most of Sweden
north to Dlr., in the east of Sweden ranging further north (to Nb.); South and Central
Finland, north to Om and Ok; in Norway only mentioned from the south-east (Ø, AK,
VE). - Widely distributed throughout most of Europe, southwards to the northern
Mediterranean, eastwards to Siberia and Central Asia.

Biology. In stagnant, eutrophic and well vegetated fresh water, mainly on open
ground, in shallow water among vegetation. Sometimes found in drift on the seashore.
March-October; larvae and pupae are found in July in stems of *Oenanthe,* adults
emerged at the end of July.

Subgenus *Methydrus* Rey, 1885

Methydrus Rey, 1885, Annls Soc. linn. Lyon 31 (1884): 253.
Type species: *Hydrophilus minutus* Fabricius (sensu Rey, 1885 [*nec Hydrophilus
minutus* Fabricius, 1775] = *Hydrophilus affinis* Thunberg, 1794), by subsequent
designation (d'Orchymont, 1939).

Maxillary palpi as in *Lumetus*. Head with inconspicuous, pronotum and elytra without (or with very few) coarser setiferous punctures. Posterior margin of last visible abdominal sternite with a small semicircular emargination in middle (Fig. 316). Claws in
♂ only slightly more strongly curved (and a little more angularly) than in ♀.

109. *Enochrus affinis* (Thunberg, 1794)
    Figs 316, 322.

Hydrophilus affinis Thunberg, 1794, Diss. Ins. Suec. 6: 73.
Enochrus minutus auctt.; *nec* (Fabricius, 1775).

3.0-3.8 mm. Head and ventral surface of body black; head at most with a pair of small, rather dark reddish preocular spots, or very seldom with more distinct (still rather small) paler preocular spots. Pronotum very dark, almost black, with reddish margins, except for middle portion of posterior margin; elytra almost uniformly reddish brown to yellowish brown, the interstice between the suture and the sutural stria normally not darker (the sutural stria itself often slightly darker). Entire dorsal surface rather finely and densely punctured. Maxillary palpi reddish, 2nd segment extensively darkened in middle; terminal segment almost uniformly dark, brownish to black, much darker than penultimate; antennae reddish yellow with darker club; legs reddish brown with paler tarsi, femora (except at apex) darker.

σ Aedeagus (Fig. 322) with long narrow median lobe, apex of parameres sharply pointed and distinctly bending outwards.

Distribution. A very common species, widespread throughout Denmark and Fennoscandia, yet from Norway only more scattered records. – Widely distributed in the Palearctic region, but in some areas (perhaps South Europe, in particular) many records no doubt apply to the closely related and recently described *isotae*.

Biology. As *ochropterus;* regularly taken in drift on the seashore. March-October.

Enochrus isotae Hebauer, 1981
    Fig. 321.

Enochrus isotae Hebauer, 1981, Ent. Bl. Biol. Syst. Käfer 77: 137.

3.0-3.7 mm. In external characters almost intermediate between *affinis* and *coarctatus;* in colour most similar to *affinis,* except that head is always provided with well defined yellowish preocular spots (which are slightly smaller than in *coarctatus*). Pronotum dark brown to blackish, with paler margins; also elytral colour as in *affinis,* i.e. the interstice between the suture and the sutural stria normally not darker. Punctuation of dorsal surface almost as in *affinis,* but on the average slightly sparser, especially on pronotum. Colour of appendages as in *affinis;* 2nd segment of maxillary palpi often a little less extensively darkened.

σ: Aedeagus (Fig. 321) with long narrow median lobe, apex of parameres more bluntly rounded than in *affinis,* not bending outwards.

Distribution. So far not recorded from Scandinavia. – The species was recently described from Yugoslavia, and has subsequently been recorded from the South of England, the South-west of France and the Netherlands. Its distribution is still not known in detail, but the species is undoubtedly widespread in Central and South Europe. So far it has not been recorded from North Germany, but it probably occurs there, and it is possible that the species may also be found in South Scandinavia.

Biology. In stagnant fresh water, preferring fairly warm, eutrophic, and well vegetated ponds, normally not in acid water. In Yugoslavia the species is found in company with *Hydrochara caraboides, Hydrophilus piceus* and some other (southern) water beetles, in Holland with *Ochthebius minimus* and dytiscids such as *Copelatus haemorrhoidalis, Hydroporus dorsalis, H. tristis* and *Bidessus unistriatus*.

110. *Enochrus coarctatus* (Gredler, 1863)
Figs 320, 323; pl. 4: 6.

Philhydrus coarctatus Gredler, 1863, Käf. Tirol: 75.

3.5-4.0 mm. Similar to the two preceeding species, but body shape on the average a little wider. Head black, with a pair of large, well defined yellowish preocular spots (Fig. 323). Pronotum normally not quite as dark as in the two preceeding species, brownish (often 4 small black spots, arranged in a square, are detectable). Elytra yellowish brown with paler margins; normally also distinctly paler medially, except for interstice between suture and sutural stria, which is normally markedly darkened (except basally). Punctuation of dorsal surface a little sparser than in *affinis*, almost as in *isotae*. Maxillary palpi paler than in both preceeding species, yellowish red with terminal segment only rather feebly darkened in middle (the apex pale); 2nd segment distinctly darkened in middle. Colour of antennae and legs as in *affinis*.

♂: Median lobe of aedeagus wide (Fig. 320).

Distribution. Common and widespread in Denmark and Sweden (except Lapland); South Finland north to Sb and Om; in Norway only mentioned from the southernmost provinces (Ø, AK, VE, AAy). – North and Central Europe, south to Central France, North Italy and the northern Balkans, eastwards to Siberia.

Biology. At the edges of stagnant fresh water, often in more or less eutrophic and well vegetated, mostly neutral, yet sometimes also acid water, often in woodland. It lives in shallow water among vegetation, in moss, etc. Sometimes found in drift on the seashore. March-October.

Genus *Cymbiodyta* Bedel, 1881

Cymbiodyta Bedel, 1881, Faune Col. bass. Seine 1: 307.
Type species: *Hydrophilus marginellus* Fabricius, 1792, by monotypy.

Similar to *Enochrus* and *Helochares*. Posterior margin of pronotum not margined. Elytra with impressed sutural stria, reaching from apex anteriorly to well before middle, otherwise without striae (in some Nearctic species, however, with a total of 10 punctured striae). Head, pronotum and elytra with groups of coarser setiferous punctures arranged as in *Helochares* and *Enochrus*. Mesosternum raised posteriorly to form a strong median dentiform process, or (in some Nearctic species) with a variable transverse ridge posteriorly. Metasternum on raised middle portion with a smooth and shining postero-medial field. Abdomen with 5 visible sternites, posterior margin of last visible sternite entire or with a small apical emargination. Maxillary palpi as in *Helochares,* terminal segment only slightly shorter than penultimate. Legs almost as in *Helochares,* except for meso- and metatarsi (Fig. 266), which are 4-segmented (with long basal segment), due to complete reduction of the small basal segment present in the related genera. Claws in ♂ a little more strongly curved than in ♀.

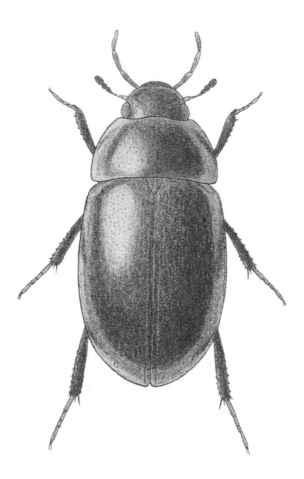

Fig. 324. *Cymbiodyta marginella* (F.), length 3.3-4.3 mm.

The genus comprises about 30 known species, all except one being restricted to the Nearctic region; the only Palearctic species occurs also in North Europe.

111. *Cymbiodyta marginella* (Fabricius, 1792)
   Figs 266, 324; pl. 4: 3.

Hydrophilus marginellus Fabricius, 1792, Ent. Syst. 1 (1): 185.

3.3-4.3 mm. Black; margins of pronotum (except middle portion of posterior margin) and lateral and apical margins of elytra rather broadly yellowish brown; pale portions of elytra with some more or less distinct longitudinal series of small blackish spots. Dorsal surface shining, with uniform, rather fine and dense punctuation; setiferous punctures of head, pronotum and elytra rather inconspicuous, on elytra very indistinct. Posterior margin of last visible abdominal sternite with a small, very feeble emargination in middle. Maxillary palpi reddish to brownish yellow; antennae yellowish red with darker club; legs piceous to black with much paler, reddish tarsi.

Distribution. Widespread and generally not uncommon, found in most parts of South Scandinavia; entire Denmark; most Swedish provinces north to Vrm.; in Finland only known from a few southern provinces (Al, Ab, N, Ka, St); in Norway only recorded from VE and Ry. – Widely distributed in Europe, ranging from Fennoscandia and the British Isles to the northern Mediterranean area, from West Europe to the Caucasus and Central Asia.

Biology. At the edges of stagnant, usually more or less eutrophic, fresh water, occurring in shallow water among vegetation, it wet moss, under leaf litter, etc. Obviously the species is fairly euryoecious, inhabiting both acid waters (e.g. *Sphagnum*-pools) and – perhaps more often – neutral or basic waters. Mainly on rather open ground, but also in woodland. Sometimes found at the slower reaches of running waters. Also in drift on the seashore. March-October.

SUBFAMILY CHAETARTHRIINAE

Body contour evenly curved, not interrupted between pronotum and elytra. Head and pronotum without distinct impressions, surface evenly convex. Mesosternum hardly raised medially, at most with a short postero-median ridge or tubercle, and raised posterior margin. Abdomen (Fig. 326) with 5 visible sternites; 1st sternite on each side with a large hyaline lobe, supported (and covered) by a fringe of long stiff yellowish setae, inserted at anterior margin of the sternite; 2nd sternite on each side with a large concavity receiving the hyaline lobes of 1st sternite; first two sternites finely and sharply ridged in middle, the other sternites not ridged, finely and densely pubescent (rest of ventral surface of body glabrous and more shining. Maxillary palpi at most slightly longer than antennae, 2nd segment often distinctly thicker than the following segments; terminal segment distinctly longer than penultimate. Antennae 9-segmented with compact club (Fig. 327). Tarsi 5-segmented, basal segment of meso- and metatarsi small, markedly shorter than 2nd segment. Tibiae without fringes of long swimming hairs.

A rather small subfamily, comprising only a little more than 50 known species and 3 genera. Only one genus occurs in North and Central Europe.

Genus *Chaetarthria* Stephens, 1832

Chaetarthria Stephens, 1832, Ill. Brit. Ent. Mandib. 5: 401.
Type species: *Hydrophilus seminulum* Herbst, 1797, by monotypy.

Head rather strongly narrowed anteriorly, eyes slightly convex, hardly protruding; anterior margin of clypeus feebly concave; labrum rather large. Pronotum short and very wide, posterior and anterior angles rather obtusely rounded and indistinct. Scutellum about as long as wide. Elytra short and wide, very strongly convex with strongly rounded sides; with a fine sharp sutural stria reaching from apex anteriorly to well before middle; elytra otherwise without striae. Prosternum anterior to procoxae reduced, very short. Procoxae large, obliquely transverse. Mesosternum not elevated in middle, only posterior margin distinctly raised. Metasternum short and wide. Maxillary palpi shorter than antennae. Antennae (Fig. 327) with large globular 2nd segment and rather compact club, whose terminal segment is smaller than the penultimate. Legs

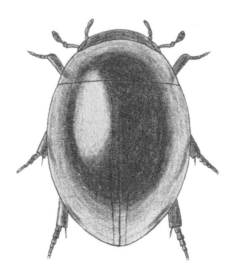

Fig. 325. *Chaetarthria seminulum* (Hbst.), length 1.3-1.7 mm. (After Victor Hansen).

210

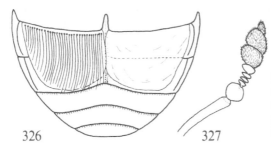

Figs 326, 327. *Chaetarthria seminulum* (Hbst.). – 326: abdomen in ventral view (long setae removed from right half); 327: right antenna.

rather short and stout. Body to some extent with the power of rolling up (as in *Agathidium* (Leiodidae), but less markedly so).

The genus comprises almost 50 known species, and is represented in the Palearctic, the Nearctic, the Neotropical and the Australian regions. Only 1 species occurs in Europe

112. *Chaetarthria seminulum* (Herbst, 1797)
    Figs 325-327; pl. 3: 5.

Hydrophilus seminulum Herbst, 1797, Natursyst. all. bek. in-u. ausl. Ins. 7: 314.

1.3-1.7 mm. Black, lateral margins of pronotum, and the elytral apex often slightly paler. Head and pronotum with even and extremely fine and dense microscopical punctuation; elytra with similar, but more obsolete microsculpture, and with sparse and rather shallow, fine punctuation, which is often indistinct antero-medially. Appendages reddish brown, often rather dark, especially the maxillary palpi.

Distribution. Rather common, found in most of Denmark and Fennoscandia north to the Arctic Circle. – Widely distributed throughout most of the Palearctic region.

Biology. At the edges of stagnant, normally eutrophic, well vegetated fresh water; mostly living in wet mud, only more seldom among vegetation in the water; occasionally also on muddy banks of slower reaches of streams. Found throughout the year, but most abundant in the spring, when it is often sieved from flood-debris, e.g. at the edges of lakes or ponds; larvae are found in July.

Subfamily Hydrophilinae

Body contour evenly curved (or almost so), hardly interrupted between pronotum and elytra. Head and pronotum without distinct impressions, the surface rather evenly convex. Mesosternum very strongly raised in middle, forming a high longitudinal keel. Metasternum markedly ridged in middle, the ridge anteriorly closely articulating to the mesosternal keel, thus forming a common sternal keel (Fig. 328), posteriorly continued in an acute or obtuse, in some forms very long spine. Abdomen with 5 visible sternites. Maxillary palpi much longer than the antennae, their 2nd segment not dilated, the terminal segment at most as long as penultimate. Antennae 9-segmented, the club loose, more or less asymmetrical. Tarsi 5-segmented, dorsal face of meso- and metatarsi with a conspicuous fringe of rather long swimming hairs. Tibiae without fringes of long swimming hairs (except for mesotibiae of some forms).

The subfamily comprises 8 genera and more than 150 known species, represented in all parts of the world. 2 genera occur in Europe.

Key to genera of Hydrophilinae

1   Length 14-18 mm. Prosternum tectiform, highly keeled in
     middle, not excavate postero-medially . . . . *Hydrochara* Berthold (p. 212)
-    Length 32-48 mm. Prosternum strongly elevated in middle,
     hood-like, with a narrow and deep postero-median excava-
     tion (Fig. 334) . . . . *Hydrophilus* Müller (p. 215)

Genus *Hydrochara* Berthold, 1827

Hydrochara Berthold, 1827, Latr. nat. Fam. Thier.: 355.
   Type species: *Dytiscus caraboides* Linnaeus, 1758, by subsequent designation (Méguignon, 1937).
Hydrophilus auctt.; *nec* Müller, 1764.

Large species. Head with large, strongly convex, protruding eyes, bordered medially by a row of coarse and close-set setiferous punctures; clypeus on each side with an arc (convex anteriorly) of similar, or even coarser punctures; anterior margin of clypeus truncate. Pronotum widest near base, moderately strongly narrowed anteriorly, at most feebly narrowed posteriorly, the posterior angles rounded. Pronotum on each side with an anterior, oblique (in middle interrupted) row of strong setiferous punctures, and a postero-lateral, very irregular group of similar punctures. Elytra moderately convex, with 10 obsolete or reduced series of extremely fine punctures, and with 5 conspicuous, more or less regular rows of coarse setiferous punctures, standing in the alternate interstices between the reduced striae. Prosternum tectiform, highly keeled medially, the keel sometimes continued posteriorly in a rather long acute spine. Metasternal ridge (Fig. 328) continued posteriorly in an acute or obtuse spine, that hardly exceeds the posterior margin of 1st visible abdominal sternite. Abdomen with dense hydrofuge pubescence, except for a small glabrous apical field on last visible sternite (only in a few exotic species without such glabrous field). Maxillary palpi long and slender, markedly longer than the antennae, terminal segment shorter than penultimate. Antennal club rather asymmetrical. Legs moderately long; profemora finely and densely punctured and pubescent basally, glabrous distally; meso- and

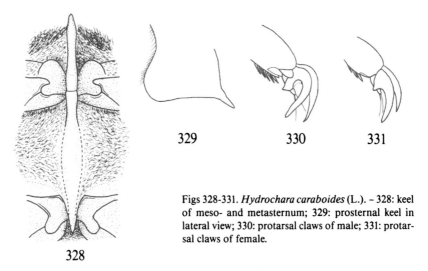

329   330   331

Figs 328-331. *Hydrochara caraboides* (L.). – 328: keel of meso- and metasternum; 329: prosternal keel in lateral view; 330: protarsal claws of male; 331: protarsal claws of female.

328

metafemora entirely glabrous. Meso- and metatarsi with a conspicuous fringe of long swimming hairs on their dorsal face. Protarsi simple in both sexes. The claws with strong basal tooth, in ♂ strongly curved at base, distally straightened (Fig. 330, in ♀ evenly curved (Fig. 331).

A genus of about 20 described species, represented in the Palearctic, the Nearctic, the Afrotropical and the Oriental regions. Only 1 species occurs in North Europe.

113. **Hydrochara caraboides** (Linnaeus, 1758)
Figs 328-332; pl. 4: 8.

Dytiscus caraboides Linnaeus, 1758, Syst. Nat. (10) 1: 411.

14-18 mm. Black, dorsal surface shining, with faint dark greenish tinge (distinct especially on live specimens), sometimes with stronger greenish or purplish lustre. Dorsal surface besides the setiferous punctures with only very fine, dense, and on elytra somewhat obsolete, punctuation; elytra in addition with extremely fine reticulate microsculpture. Prosternal keel continued posteriorly in a rather long acute spine (Fig. 329). Maxillary palpi reddish brown; antennae yellowish red with piceous club; legs piceous club; legs piceous to black, occasionally partially paler; very seldom the legs may be pale yellowish brown, with darker tarsi and slightly darker femora.

Distribution. In Scandinavia only found in the south; rather common in eastern Denmark (SJ, F, LFM, SZ, NWZ, NEZ, B); otherwise rather local and, towards the north rare; found in most Swedish provinces north to Vrm. and Hls.; in Finland only found in Al and Ta; in Norway only Ø and Bø. – Widely distributed throughout most of the Palearctic region, from Fennoscandia and the British Isles to the Mediterranean, exclusive of the Iberian Peninsula (except northernmost Spain), from West Europe to Siberia.

Biology. In stagnant, eutrophic fresh water with grassy and muddy bottom, especially on open ground, sometimes rather abundant; usually occurring on very shallow water among vegetation. The eggs are laid in spring or early summer; larvae are found from May to July. The adults normally hatch in summer, although they may occasionally emerge already in May. They are mainly found in spring and late summer or autumn. Sometimes also taken in drift on the seashore, or at light.

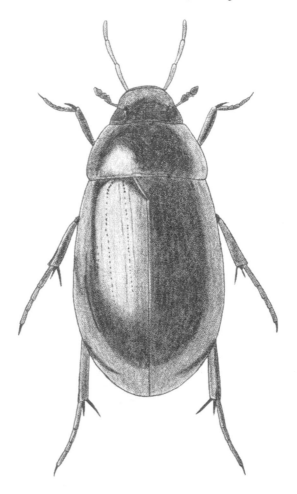

Fig. 332. *Hydrochara caraboides* (L.), length 14-18 mm. (After Victor Hansen).

Genus *Hydrophilus* Müller, 1764

Hydrophilus Müller, 1764, Fauna Ins. Fridr.: 16.
  Type species: *Dytiscus piceus* Linnaeus, 1758, by subsequent designation (Latreille, 1810).
Hydrous Linnaeus, 1775, Diss. Ent.: 7.
  Type species: *Dytiscus piceus* Linnaeus, 1758 (see below).

Very large species. Body shape more oval than in *Hydrochara,* elytra more strongly narrowed posteriorly. Head and pronotum almost as in *Hydrochara,* but anterior margin of clypeus broadly and feebly emarginate, exposing articulation membrane of labrum. Group of setiferous punctures of head and pronotum arranged as in *Hydrochara,* but less conspicuous, especially those of pronotum; also the elytral rows of setiferous punctures finer than in *Hydrochara;* the 10 obsolete series of extremely fine punctures often visible only laterally and apically. Exterior apical angles of elytra

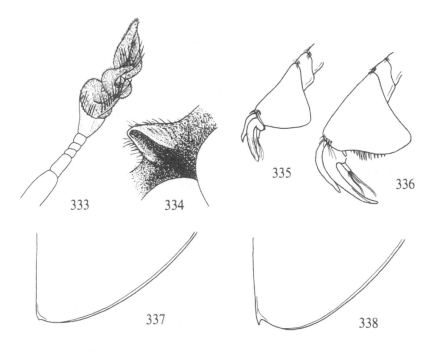

Figs 333, 334. *Hydrophilus aterrimus* Eschtz. - 333: right antenna; 334: prosternum in ventro-lateral view.
Figs 335, 336. Distal part of protarsus of *Hydrophilus*-males. - 335: *aterrimus* Eschtz.; 336: *piceus* (L.).
Figs 337, 338. Apex of right elytron of *Hydrophilus.* - 337: *aterrimus* Eschtz.; 338: *piceus* (L.).

215

obtusely rounded. Prosternum (Fig. 334) strongly raised in middle, hood-like, narrowly and deeply excavate postero-medially, to receive apex of mesosternal keel. Posterior acute spine of metasternal ridge longer than in *Hydrochara*, distinctly exceeding posterior margin of 1st visible abdominal sternite (in some exotic forms very long). Mesosternal keel variably furrowed or grooved in middle (strongest in ♂♂). Abdominal sternites with fine and dense hydrofuge pubescence; in some species (including the two treated here) the pubescence is restricted to the 1st visible sternite and the lateral margins of the following sternites, in others covering the entire ventral surface of abdomen. Maxillary palpi as in *Hydrochara;* the antennal club very asymmetrical, besides the hydrofuge pubescence with some long fine setae (Fig. 333) Legs almost as in *Hydrochara*, except for protarsi of ♂♂, which are distinctly dilated, with the claw segment (in the here treated species) forming a large triangular plate, and the claws which are enlarged and strongly curved at base (Figs 335, 336).

The name *Hydrous* was introduced by Linnaeus (1775) merely as a replacement for *Hydrophilus,* the latter name being however validated in 1764 by O. F. Müller. Thus, *Hydrous* is a junior objective synonym of *Hydrophilus,* having the same type-species, and the two names are indissolubly linked (For details see Pope, 1985).

A genus with a world wide distribution, comprising about 50 known species; 2 of these occur in North and Central Europe.

Key to species of *Hydrophilus*

1   All abdominal sternites sharply tectiform, almost ridged in
    middle. Elytral apex at sutural angle with a small acute spine
    (Fig. 338) . . . . 114. *piceus* (Linnaeus)
–   Abdominal sternites (except the last visible) more bluntly rais-
    ed medially, not sharply tectiform. Elytral apex without small
    acute spine at sutural angle (Fig. 337) . . . . 115. *aterrimus* Eschscholtz

114. ***Hydrophilus piceus*** (Linnaeus, 1758)
    Figs 336, 338.

Dytiscus piceus Linnaeus, 1758, Syst. Nat. (10) 1: 411.

34-48 mm. Black, dorsal surface with faint dark greenish tinge (distinct especially on live specimens); sometimes with stronger, dark greenish or purplish hue, particularly towards the lateral margins of elytra. Entire dorsal surface shining, without distinct punctuation besides the setiferous punctures, except for an extremely fine micropunctuation. Elytral apex at sutural angle with a small acute spine (Fig. 338). Mesosternal keel in middle with a rather short furrow, that does not reach the anterior apex of the keel. Abdominal sternites sharply tectiform, almost ridged medially. Maxillary palpi and antennae reddish; legs piceous to black, tarsi often brownish.

    ♂: Triangular protarsal claw segment large (Fig. 336), the inner apex reddish. Mesosternal keel wider than in ♀.

Distribution. A comparatively rare and local species; in Denmark found sparsely in SJ, EJ, F, LFM, SZ, NWZ and NEZ; in Sweden known only from Sk., Öl., Gtl., Ög. and Upl.; in Finland only recorded from N and Sa; in Norway only AK. – Widely dis-

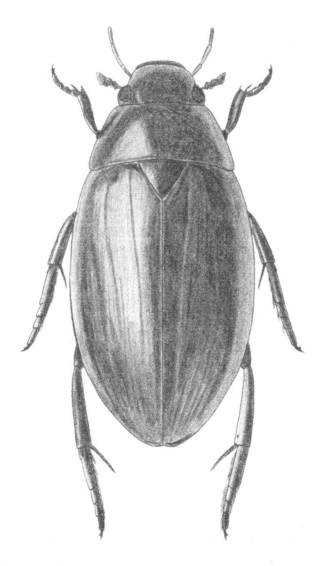

Fig. 339. *Hydrophilus aterrimus* Eschtz., length 32-43 mm. (After Victor Hansen).

tributed throughout the Palearctic region, ranging from South Scandinavia to the Mediterranean, from West Europe to North India and Siberia.

Biology. In stagnant, rather eutrophic, and well vegetated, clear fresh water with muddy bottom, mainly in rather sunny ponds. Adults feed among vegetation (e.g. *Lemna*) on plant material just below the water surface, often on rather shallow water. The eggs are laid in May-June in silk cocoons, containing normally up to about 50-70 eggs each. The larvae feed predominantly on snails (mainly *Limnaea* spp.), and are full-grown in about 4 weeks; pupation takes place in humid soil near the edge of the water. The adults emerge in July-August, and may live for 2-3 (perhaps more) years. They are active flyers, regularly coming to light.

115. *Hydrophilus aterrimus* Eschscholtz, 1822
   Figs 333-335, 337, 339; pl. 4: 7.

Hydrophilus aterrimus Eschscholtz, 1822, Ent. 1: 128.

32-43 mm. Very similar to *piceus*, but differs by the medially more bluntly raised abdominal sternites, which are not sharply tectiform (except for the last sternite), absence of spine at sutural angle of elytra (Fig. 337), and longer median furrow of mesosternal keel, the furrow reaching almost to the anterior apex of the keel. The elytra are normally less strongly narrowed posteriorly than in *piceus*.

♂: Triangular protarsal claw segment smaller than in *piceus* (Fig. 335), uniformly piceous to black. Mesosternal keel not wider than in ♀ (cf. *piceus*-♂).

Distribution. Relatively rare and very local; in Denmark scattered records from EJ, WJ, F, LFM, NWZ, NEZ, and B; in Sweden only Sk., Bl., Hall., Öl. and Gtl.; from Finland only a few specimens from N; not found in Norway. – North and Central Europe (not the British Isles), south to the northern Mediterranean; eastwards to Siberia and Transcaspia.

Biology. Habitat and life history as in *piceus;* the adults are only seldom taken at light, and seem to fly only occasionally.

SUBFAMILY BEROSINAE

Body contour evenly curved, or slightly interrupted between pronotum and elytra. Head and pronotum without distinct impressions, their surface rather evenly convex. Scutellum markedly longer than wide. Mesosternum often ridged in middle, sometimes strongly raised. Metasternum with raised middle portion, but only seldom (in some exotic forms) ridged; meso- and metasternum never forming a continuous ridge. Abdomen normally with 5 visible sternites, of which the 5th is variably excised posteriorly, so a 6th, more or less retractable sternite is exposed (in some exotic forms with only 4 visible abdominal sternites, and without excised sternite). Maxillary palpi as long as, or a little longer than antennae, their terminal segment longer than penultimate; 2nd segment not dilated. Antennae 7- or 8-segmented, with loose club. Tarsi 5-segmented, protarsi of ♂♂ (except of some exotic genera) 4-segmented; basal segment of meso- and metatarsi minute. Meso- and metatibiae and -tarsi with conspicuous fringes of long fine swimming hairs (Fig. 340).

A rather large subfamily with a world wide distribution, represented mainly in the warmer climates. It comprises 5 genera and almost 300 known species. Only one genus occurs in Europe.

Genus *Berosus* Leach, 1817

Berosus Leach, 1817, Zool. Misc. 3: 92.
  Type species: *Dytiscus luridus* Linnaeus, 1761, by monotypy.

Eyes large, strongly convex and rather strongly protruding. Pronotum widest near base, only moderately strongly narrowed anteriorly, the posterior angles rounded. Elytra often rather strongly convex, widest at or behind middle. Body contour not forming a continuous curve, but feebly interrupted between pronotum and elytra. Elytra punctato-striate, each with 10 complete striae and a short intercalary stria at base between 1st and 2nd stria; 6th and 7th striae not reaching base (in some exotic species the striae may be more obsolete or even indistinct, with their punctures not differing from the punctures of the interstices). Dorsal surface almost glabrous (with only few sparse setae) or – in some exotic species – distinctly pubescent. Ventral surface of body dull, finely and very densely punctured, with dense hydrofuge pubescence. Prosternum anterior to procoxae reduced, very short, ridged medially. Mesosternum large, ridged in middle and often more strongly raised posteriorly, to form a narrow more or less acute process. Metasternum raised medially, the middle portion flattened, sharply angularly prolonged posteriorly, with a distinct longitudinal furrow. Abdomen with 5 visible sternites, posterior margin of 5th sternite excised, exposing a 6th, more or less retractable sternite (Fig. 344). Antennae 7-segmented. Legs rather long and slender, femora at least basally with very fine and dense punctuation and pubescence, distally glabrous and shining. Tarsi 5-segmented, except for protarsi in ♂♂, which are 4-segmented with the two basal segments markedly dilated (Fig. 341).

A large genus, comprising more than 200 known species, and represented in all major zoogeographical regions. Only 3 species occur in Fennoscandia and Denmark.

Key to species of *Berosus*

1   Each elytron with a long acute latero-apical spine, and
     sharply angular sutural angle (Figs 342, 343) .... 116. *spinosus* (Steven)
–   Elytra without apical spines, sutural angles rounded 2
2 (1) Pronotum densely and rather strongly punctured, distance

between the punctures generally smaller than the punctual
diameter. Elytral interstices rather coarsely and uniformly
punctured .... 118. *luridus* (Linnaeus)
- Pronotum much more finely punctured, distance between
the punctures generally larger than the punctual diameter.
Elytral interstices rather finely punctured, the alternate interstices with rows of markedly coarser setiferous punc-
tures .... 117. *signaticollis* (Charpentier)

Subgenus *Enoplurus* Hope, 1838

Enoplurus Hope, 1838, Col. Man. 2: 128.
Type species: *Hydrophilus spinosus* Steven, 1808, by present designation.

Each elytron with a long acute latero-apical spine; the sutural angle sharply angular,
often dentiform (Figs 342, 343).

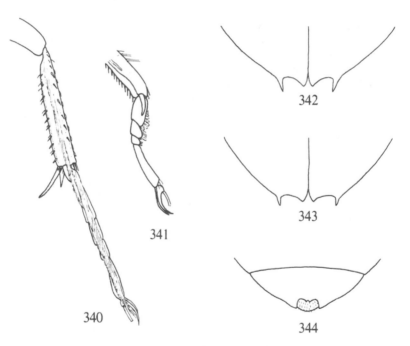

Figs 340-343. *Berosus spinosus* (Stev.). - 340: mesotibia and -tarsus; 341: protarsus of male; 342: elytral apex of female; 343: elytral apex of male.
Fig. 344. Last abdominal sternites of *Berosus signaticollis* (Charp.).

220

In Hope's description of *Enoplurus,* it was mentioned that it includes several species, but only two species were mentioned by name, viz. *Berosus spinosus* (Stev.) and *Berosus orientalis* Hope. The latter was designated as the type-species by Hope, but in fact, Hope never described any species by that name (in fact no *Berosus* has been described under the name *orientalis*), so Hope's own designation must be considered invalid. Thus, the only other species mentioned by name by Hope, viz. *B. spinosus,* will be the type-species.

116. *Berosus spinosus* (Steven, 1808)
      Figs 340-343; pl. 4: 4.

Hydrophilus spinosus Steven, 1808, *in* Schoenh. Syn. Ins. 2: 8.

4.5-5.7 mm. Dorsal surface brownish yellow, head often variably darkened posteriorly; elytra often with some obsolete, slightly darker spots. Ventral surface of body piceous to black. Head rather finely and not very densely punctured (the punctuation a little stronger and denser posteriorly); pronotum a little finer and more sparsely punctured, especially medially. Elytral striae fine, rather sharply impressed, weaker anteriorly, the strial punctures generally rather fine. Interstices flat or almost so, finely and somewhat sparsely punctured, 3rd and 5th interstices each with a more or less regular row of coarser setiferous punctures; punctuation of the interstices becoming gradually much finer (sometimes almost indistinct) laterally and apically. Mesosternum rather evenly ridged throughout, not more strongly raised posteriorly. Appendages yellowish red; apex of maxillary palpi darker.
      ♂: Elytra shining, without microsculpture. Elytral apex, Fig. 343.
      ♀: Elytra somewhat duller, except anteromedially, with fine but distinct reticulate microsculpture. Pronotum often with similar, but more obsolete microsculpture. Sutural angle of elytra a little sharper angular than in ♂ (Fig. 342).

Distribution. A relatively rare species, found locally along the coasts of southern Scandinavia; in Denmark found in most provinces; in Sweden only known from the south (Sk., Bl., Hall., Sm., Öl., Gtl.) and from Sdm. and Upl.; in Norway only recorded from Ø and AK; in Finland only Al, Ab and N. – Widespread along the European coasts, from Fennoscandia to the Atlantic coast of West Europe; southwards to the Mediterranean; also round the Black Sea and the Caspian Sea and on saline inland localities in Central and SE. Europe and South Russia.

Biology. A halobiontic species, living exclusively on salt marshes, in brackish, often shallow and sparsely vegetated pools; often fairly abundant where it lives. In Scandinavia found exclusively along the coasts, but farther south also on saline inland localities; only accidentally in fresh water. The species may easily be taken for a dytiscid, when it swims rapidly over the bottom. It is also an active flyer, and may be taken at light. May-September; larvae and pupae are found at the beginning of July.

Subgenus *Berosus* Leach, 1817, s.str.

Berosus Leach, 1817, Zool. Misc. 3: 92.
Type species: *Dytiscus luridus* Linnaeus, 1761, by monotypy.

Differs from *Enoplurus* by the absence of apical elytral spines, and the rounded sutural angles.

117. *Berosus signaticollis* (Charpentier, 1825)
   Fig. 344.

Hydrophilus signaticollis Charpentier, 1825, Horae Ent.: 204.

4.8-6.0 mm. Dorsal surface brownish yellow; head black with bright greenish, bronze, purplish or sometimes bluish metallic lustre. Pronotum with a large elongate black spot in middle (with metallic reflections as on head); this median spot often narrowly divided by a pale median stripe. Elytra usually with some obsolete, slightly darker spots. Ventral surface of body piceous to black. Head rather finely and rather densely punctured; middle portion of pronotum slightly finer and a little less densely punctured (midline impunctate), the punctuation becoming gradually stronger laterally. Elytral striae rather fine, sharply impressed, the strial punctures fine or very fine, becoming gradually distinctly stronger laterally, apically and near base. Interstices very weakly convex, finely and somewhat sparsely punctured; 3rd, 5th and 7th interstices each with a more or less regular row of markedly coarser (on interstice 7 very sparse) punctures. Mesosternal ridge strongly raised posteriorly, to form a narrow, acute dentiform process. Appendages yellowish red, apex of maxillary palpi and femora basally a little darker.
   ♂: Elytra shining, without microsculpture.
   ♀: Elytra rather dull, with fine, but distinct reticulate microsculpture.

Distribution. A rare to very rare species in southernmost Scandinavia; in Denmark found only very sparsely in SJ, EJ, WJ and NEJ; in Sweden only Sk., Bl., Hall., Öl., Gtl. and Vg.; not in Norway and Finland. – Europe, ranging from the British Isles and South Scandinavia to the Mediterranean (including North Africa), from West Europe to Siberia.

Biology. In stagnant fresh water, mainly (or perhaps exclusively) in clear, oligotrophic and acid, sparsely vegetated and rather shallow, sometimes temporary pools; in Denmark typically on moors.

Note. Hellén (1939) gives a doubtful record from Norway (Ø/AK). It has not been possible to confirm this record.

118. *Berosus luridus* (Linnaeus, 1761)
   Fig. 345.

222

Dytiscus luridus Linnaeus, 1761, Fauna Suec. (2): 214.

3.5-4.6 mm. Colour as in *signaticollis,* except that dark middle spot of pronotum is normally much larger. Head and pronotum densely and rather strongly punctured (stronger than in *signaticollis*). Midline of pronotum forming a variably narrow impunctate stripe. Elytral striae rather strongly punctured, the strial punctures becoming distinctly stronger laterally. Interstices more strongly and densely punctured than in *signaticollis,* but without stronger punctures in 3rd, 5th and 7th interstices; punctua-

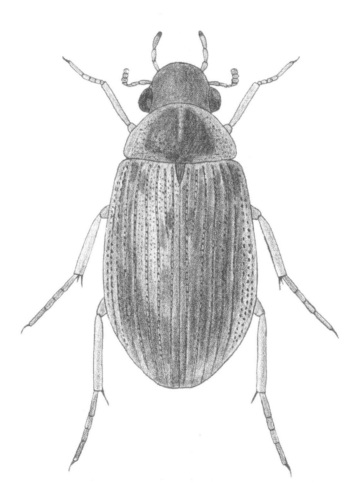

Fig. 345. *Berosus luridus* (L.), length 3.5-4.6 mm. (After Victor Hansen).

223

tion of interstices becoming gradually finer laterally and apically (so on these portions it is no stronger, or even finer, than in *signaticollis*). Mesosternum almost as in *signaticollis,* but with more bluntly rounded apex of the mesosternal process.

♂: Elytra shining, without microsculpture.

♀: Elytra a little duller, with very fine, rather obsolete, reticulate microsculpture.

Distribution. Fairly common and widespread throughout most of Denmark and Fennoscandia north to about 68°N. – Widely distributed throughout the Palearctic region.

Biology. In stagnant fresh water, mainly in rather eutrophic, open, shallow pools or ponds, with grassy and somewhat clayey bottom; occasionally found also in brackish water. The species is sometimes found in drift on the seashore, or at light. March-October.

Plate 1

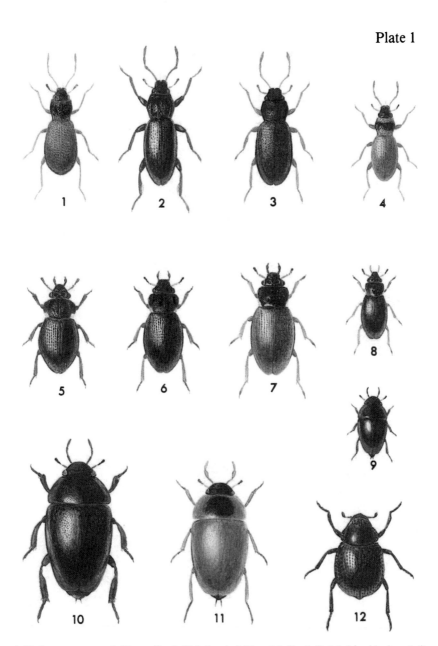

1: Hydraena testacea; 2: H. gracilis; 3: H. britteni; 4: H. pulchella; 5: Ochthebius bicolon; 6: O. minimus; 7: O. marimus; 8: O. evanescens; 9: Limnebius aluta; 10: L. truncatellus; 11: L. pappo-sus; 12: Georissus crenulatus.

Plate 2

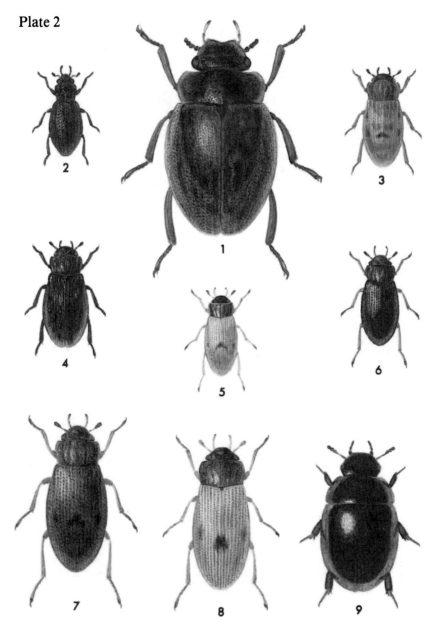

1: Spercheus emarginatus; 2: Hydrochus brevis; 3: Helophorus nubilus; 4: H. tuberculatus; 5: H. minutus; 6: H. flavipes; 7: H. aequalis; 8: H. pallidus; 9: Coelostoma orbiculare.

Plate 3

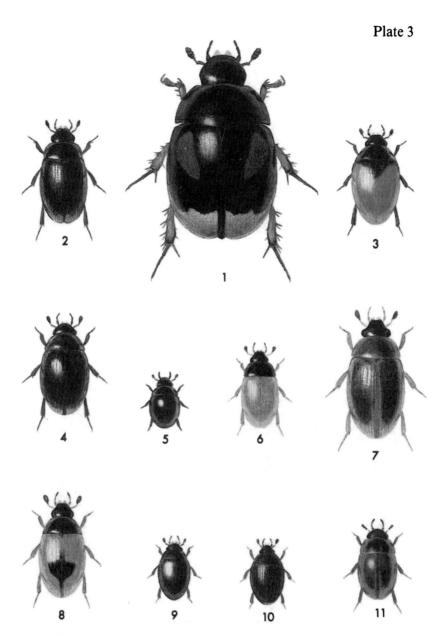

1: Sphaeridium scarabaeoides; 2: Cercyon littoralis; 3: C. melanocephalus; 4: C. marinus; 5: Chaetarthria seminulum; 6: Cercyon quisquilius; 7: C. laminatus; 8: C. unipunctatus; 9: Megasternum obscurum; 10: Cryptopleurum minutum; 11: Anacaena limbata.

Plate 4

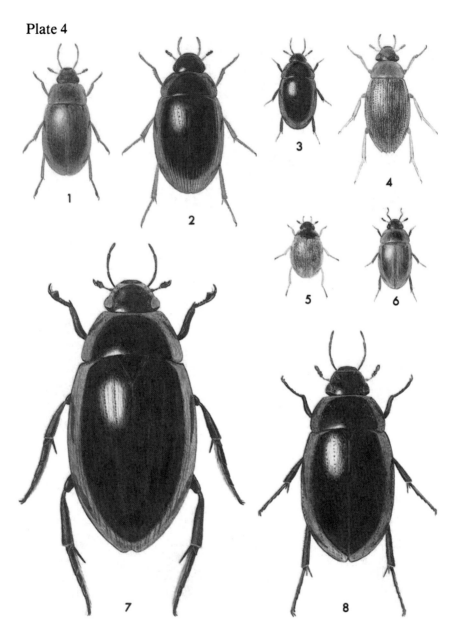

1: Helochares obscurus; 2: Hydrobius fuscipes; 3: Cymbiodyta marginella; 4: Berosus spinosus;
5: Laccobius minutus; 6: Enochrus coarctatus; 7: Hydrophilus aterrimus; 8: Hydrochara caraboides.

Catalogue

		Germany	G. Britain	SJ	EJ	WJ	NWJ	NEJ	F	LFM	SZ	NWZ	NEZ	B	Sk.	Bl.
					DENMARK											
Ochthebius gibbosus Germ.		●														
O. auriculatus Rey	1	●	●	●	●	●		●	●						●	
O. dilatatus Stph.	2	●	●	●	●	●		●	●	●	●	●	●		●	
O. bicolon Germ.	3	●	●	●	●	●		●	●	●	●	●	●	●	●	●
O. stockmanni B.-Br.	4															
O. kaninensis Popp.	5															
O. exaratus Muls.		●														
O. narentinus Reitt.		●														
O. minimus (F.)	6	●	●	●	●	●	●	●	●	●	●	●	●	●	●	●
O. nilssoni Hebauer	7															
O. marinus (Payk.)	8	●	●	●	●	●	●	●	●	●	●	●	●		●	●
O. lenensis Popp.	9	●														
O. viridis Peyr.	10	●	●			●	●		●	●	●	●			●	
O. pusillus Stph.		●	●													
O. evanescens J. Sahlbg.	11															
Hydraena testacea Curt.	12	●	●	●		●			●							
H. palustris Er.	13	●	●	●	●	●		●	●	●	●					
H. britteni Joy	14	●	●	●	●	●		●	●	●	●	●	●	●	●	●
H. riparia Kugel.	15	●	●	●	●	●		●	●	●	●	●	●	●	●	●
H. bohemica Hrbáček		●														
H. sternalis Rey		●														
H. nigrita Germ.	16	●	●	●						●			●	●		
H. pulchella Germ.	17	●	●	●	●			●	●				●	●		
H. gracilis Germ.	18	●	●	●	●	●		●	●		●		●	●		
H. belgica d'Orch.		●														
Limnebius truncatellus (Thbg.)	19	●	●	●	●	●	●	●	●	●	●	●	●	●	●	●
L. papposus Muls.	20	●	●	●	●	●		●	●	●	●	●	●			
L. crinifer Rey	21	●	●	●	●	●	●									
L. truncatulus Thoms.	22	●		●		●	●	●	●	●	●	●	●	●		
L. nitidus (Marsh.)	23	●	●	●					●			●	●			
L. aluta Bedel	24	●	●	●		●			●	●	●	●	●	●	●	●
L. atomus (Dft.)		●														
Spercheus emarginatus (Schall.)	25	●	●	●	●	●	●	●	●	●	●		●		●	
Hydrochus elongatus (Schall.)	26	●	●	●	●	●	●	●	●	●	●		●		●	●
H. ignicollis Motsch.	27	●	●	●	●	●	●	●	●	●	●	●	●		●	●
H. carinatus Germ.	28	●	●	●	●	●	●	●	●	●	●	●	●		●	●
H. brevis (Hbst.)	29	●	●	●	●	●	●	●	●	●	●	●	●		●	●
Georissus crenulatus (Rossi)	30	●	●	●	●	●	●	●	●	●	●	●	●		●	●

226

	Hall.	Sm.	Öl.	Gtl.	G. Sand.	Ög.	Vg.	Boh.	Dlsl.	Nrk.	Sdm.	Upl.	Vstm.	Vrm.	Dlr.	Gstr.	Hls.	Med.	Hrj.	Jmt.	Ång.	Vb.	Nb.	Ås. Lpm.	Ly. Lpm.	P. Lpm.	Lu. Lpm.	T. Lpm.
1	●							●																				
2	●																											
3	●	●	●	●		●	●		●	●	●	●	●															
4																												
5																												
6	●	●	●	●	●	●	●		●	●	●	●	●	●			●	●		●	●	●	●	●				
7																					●							
8	●	●	●	●	●		●	●			●	●					●				●	●	●					
9																												
10			●																									
11																												
12		●					●																					
13	●	●	●	●		●	●			●	●	●		●														
14	●	●	●	●		●	●	●	●	●	●	●	●	●	●	●	●	●	●	●	●	●	●	●	●	●	●	●
15	●	●	●	●	●	●	●	●	●	●	●	●	●	●	●		●	●		●								
16																												
17	●											●																
18	●	●		●		●	●	●	●	●	●	●	●	●	●					●	●	●	●	●			●	
19	●	●	●	●	●	●	●	●	●	●	●	●	●	●	●	●	●	●	●	●	●	●	●				●	●
20								●																				
21		●	●	●	●		●	●		●		●	●		●													
22		●	●	●	●		●	●		●	●	●		●	●													
23																												
24	●	●	●	●	●	●	●	●			●	●	●			●												
25			●			●	●			●		●	●			●												
26		●					●				●	●																
27	●	●	●	●		●	●	●	●	●	●	●	●	●	●	?	●	●			●							
28	●	●	●			●	●	●	●	●	●	●	●	●														
29	●	●	●	●		●	●	●	●	●	●	●	●	●	●		●	●		●	●	●	●				●	
30	●	●	●	●	●	●	●	●	●	●	●	●	●	●	●						●	●						

227

		Ø+AK	HE (s+n)	O (s+n)	B (ø+v)	VE	TE (y+i)	AA (y+i)	VA (y+i)	R (y+i)	HO (y+i)	SF (y+i)	MR (y+i)	ST (y+i)	NT (y+i)	Ns (y+i)
Ochthebius gibbosus Germ.																
O. auriculatus Rey	1															
O. dilatatus Stph.	2															
O. bicolon Germ.	3	●			●											
O. stockmanni B.-Br.	4															
O. kaninensis Popp.	5															
O. exaratus Muls.																
O. narentinus Reitt.																
O. minimus (F.)	6	●			●	●			◖				●		◗	●
O. nilssoni Hebauer	7															
O. marinus (Payk.)	8	●			●				◖							
O. lenensis Popp.	9															◗
O. viridis Peyr.	10															
O. pusillus Stph.																
O. evanescens J. Sahlbg.	11															
Hydraena testacea Curt.	12															
H. palustris Er.	13	◗	◖	◖			◗									
H. britteni Joy	14	●◖		◗◖	●	◗◖	◖	◖	◖		◗	◗	◗	◗	◗	◗
H. riparia Kugel.	15	●◖		◖	●	?	◖	?					●	◗		
H. bohemica Hrbáček																
H. sternalis Rey																
H. nigrita Germ.	16	◗					◖		◖							
H. pulchella Germ.	17															
H. gracilis Germ.	18	●◖	◖		●		●◖		◖				◖	●	◗	
H. belgica d'Orch.																
Limnebius truncatellus (Thbg.)	19	●	●	●	◖	●	●◖	◖	●◖				◗	●	●	●
L. papposus Muls.	20															
L. crinifer Rey	21															
L. truncatulus Thoms.	22	◗		●	●				◖							
L. nitidus (Marsh.)	23															
L. aluta Bedel	24	●			◖											
L. atomus (Dft.)																
Spercheus emarginatus (Schall.)	25															
Hydrochus elongatus (Schall.)	26															
H. ignicollis Motsch.	27	●		●	◖	●		◖	◖							
H. carinatus Germ.	28															
H. brevis (Hbst.)	29	●	◖		◖	●		◖								
Georissus crenulatus (Rossi)	30	●						◖								

#	Nn (ø+v)	TR (y+i)	F (v+i)	F (n+ø)	Al	Ab	N	Ka	St	Ta	Sa	Oa	Tb	Sb	Kb	Om	Ok	ObS	ObN	Ks	LkW	LkE	Le	Li	Vib	Kr	Lr
1																											
2																											
3			◐		●																						
4						●	●																				
5																				●							
6				◑◐	●	●	●	●	●	●	●	●	●	●	●	●	●	●	●	●	●			●	●	●	●
7																											
8					●	●	●	●		●		●			●											●	●
9	◀																								●	●	
10																											
11																										●	●
12																											
13						●			●				●	●												●	
14				▶	●	●	●	●	●	●	●	●	●	●	●	●	●	●	●	●	●	●	●		●	●	●
15					●	●	●	●	●	●	●	●	●	●	●										●	●	●
16																											
17						●	●		●	●		●					●									●	
18				◐			●		●	●			●	●				●	●	●					●	●	
19	◐		◐		●	●	●	●	●	●	●	●	●	●	●										●	●	
20																											
21						●			●				●	●												●	
22					●	●	●	●	●	●	●	●	●	●											●	●	
23						●																					
24					●	●	●	●		●		●													●		
25																									●	●	●
26																											
27					●	●	●	●	●	●	●	●	●	●											●	●	
28						●			●																		
29					●	●	●	●	●	●	●	●	●	●	●	●	●		●						●	●	●
30					●	●	●		●																●	●	

	Germany	G. Britain	SJ	EJ	WJ	NWJ	NEJ	F	LFM	SZ	NWZ	NEZ	B	Sk.	Bl.
Helophorus rufipes (Bosc)		●													
H. porculus Bedel	●	●													
H. nubilus F.	31 ●	●	●	●	●	●	●	●	●	●	●	●	●	●	●
H. tuberculatus Gyll.	32 ●	●	●	●	●			●	●	●	●		●	●	●
H. sibiricus (Motsch.)	33														
H. aquaticus (L.)	34 ●			●	●			●	●	●	●		●		
H. aequalis Thoms.	35 ●	●	●	●	●	●	●	●	●	●	●	●	●	●	●
H. grandis Ill.	36 ●	●	●	●	●	●	●	●	●	●	●	●	●	●	●
H. strandi Angus	37														
H. brevipalpis Bedel	38 ●	●	●	●	●	●	●	●	●	●	●	●	●	●	●
H. arvernicus Muls.	39 ●	●	●	●											
H. glacialis Villa	40														
H. granularis (L.)	41 ●	●	●	●	●	●	●	●	●	●	●		●	●	●
H. discrepans Rey	42														
H. minutus F.	43 ●	●	●	●	●	●	●	●	●	●	●	●	●		
H. lapponicus Thoms.	44							●							
H. fulgidicollis Motsch.	45 ●	●	●		●	●	●	●	●	●					
H. griseus Hbst.	46 ●	●	●	●	●	●	●	●	●	●	●	●	●	●	
H. longitarsis Woll.		●	●												
H. nanus Sturm	47 ●	●	●	●	●	●	●	●	●	●	●	●	●		
H. redtenbacheri Kuw.	48 ●									●		●			
H. pumilio Er.	49 ●							●							
H. pallidus Gebl.	50														
H. laticollis Thoms.	51 ●	●	●		●		●							●	●
H. strigifrons Thoms.	52 ●	●	●	●	●	●	●	●	●	●	●	●	●	●	●
H. croaticus Kuw.		●													
H. dorsalis (Marsh.)		●	●												
H. asperatus Rey	53 ●		●		●	●							●	●	
H. flavipes F.	54 ●	●	●	●	●	●	●	●	●	●		●			●
H. obscurus Muls.	55 ●	●	●	●	●	●	●	●	●	●	●	●	●	●	●
Coelostoma orbiculare (F.)	56 ●	●	●	●	●	●	●	●	●	●	●	●	●	●	●
Dactylosternum abdominale (F.)	●														
Sphaeridium bipustulatum F.	57 ●	●	●	●	●	●	●	●	●	●	●	●		●	●
S. lunatum F.	58 ●	●	●	●	●	●	●	●	●	●	●	●	●	●	●
S. scarabaeoides (L.)	59 ●	●	●	●	●	●	●	●	●	●	●	●	●	●	●
Cercyon ustulatus (Preyssl.)	60 ●	●	●	●	●	●	●	●	●	●	●	●	●	●	●
C. laminatus Sharp	61 ●	●						●	●	●		●		●	
C. littoralis (Gyll.)	62 ●	●	●	●	●	●	●	●	●	●	●	●	●	●	●

	Hall.	Sm.	Öl.	Gtl.	G. Sand.	Ög.	Vg.	Boh.	Dls.	Nrk.	Sdm.	Upl.	Vstm.	Vrm.	Dlr.	Gstr.	Hls.	Med.	Hrj.	Jmt.	Ång.	Vb.	Nb.	Ås. Lpm.	Ly. Lpm.	P. Lpm.	Lu. Lpm.	T. Lpm.
31	●	●	●	●	●	●	●	●			●	●	●	●	●													
32	●	●	●	●	●	●	●	●			●	●			●					●	●				●		●	
33																			●			●				●	●	●
34																												
35	●	●	●	●		●	●	●	●		●	●	●															
36	●	●	●	●		●	●	●	●	●	●	●	●	●		●				●	●	●						
37																						●				●	●	●
38	●	●	●	●	●	●	●	●	●	●	●	●	●	●						●	●	●	●				●	
39																												
40															●					●	●		●	●	●	●	●	●
41	●	●	●	●		●	●	●	●	●	●	●	●	●	●	●	●	●	●	●	●	●	●		●		●	
42																												
43	●	●	●	●	●	●	●	●	●		●	●		●	●													
44		●													●				●	●	●	●	●	●	●	●	●	●
45																												
46		●	●		●									●														
47		●	●	●		●				●	●	●	●									●						●
48																												
49																												
50																			●	●	●	●	●	●	●	●		
51	●	●	●			●	●	●			●	●		●		●				●		●	●	●	●	●	●	
52	●	●	●	●		●	●	●	●	●	●	●	●	●	●	●	●				●	●	●	●	●		●	●
53																												
54	●	●	●	●		●	●	●	●	●	●	●	●	●	●	●	●	●	●	●	●	●	●					●
55	●		●	●		●	●																					
56	●	●	●	●		●	●	●	●	●	●	●	●	●	●	●	●	●	●	●	●	●						
57	●	●	●	●		●	●	●	●	●	●	●	●	●	●	●	●	●	●	●	●		●		●		●	
58	●	●	●	●		●	●	●	●	●	●	●	●	●	●	●	●	●	●	●	●	●	●			●	●	●
59	●	●	●	●		●	●	●	●	●	●	●	●	●	●	●	●	●	●		●	●	●		●		●	
60	●	●	●	●		●	●	●	●	●	●	●	●	●														
61			●																				●					
62	●	●	●	●	●	●	●	●	●		●	●																

		Ø+AK	HE (s+n)	O (s+n)	B (ø+v)	VE	TE (y+i)	AA (y+i)	VA (y+i)	R (y+i)	HO (y+i)	SF (y+i)	MR (y+i)	ST (y+i)	NT (y+i)	Ns (y+i)
Helophorus rufipes (Bosc)																
H. porculus Bedel																
H. nubilus F.	31	●														
H. tuberculatus Gyll.	32	◐				◐										
H. sibiricus (Motsch.)	33			◐	◐											
H. aquaticus (L.)	34															
H. aequalis Thoms.	35	●			◐	●	●				●	◐	◐	●	●	●
H. grandis Ill.	36	●			◐	◐										
H. strandi Angus	37			◐									◐			
H. brevipalpis Bedel	38	●	◐	?	◐	●	●				●	◐	◐	●		◐
H. arvernicus Muls.	39															
H. glacialis Villa	40		◐	●	●			◐			◐	●	◐		◐	◐
H. granularis (L.)	41	●	●	◐	◐		●	●	◐		◐			◐	◐	
H. discrepans Rey	42															
H. minutus F.	43	●		◐	◐	●										
H. lapponicus Thoms.	44		◐	◐										◐		
H. fulgidicollis Motsch.	45															
H. griseus Hbst.	46	●			◐											
H. longitarsis Woll.																
H. nanus Sturm	47				◐											
H. redtenbacheri Kuw.	48															
H. pumilio Er.	49															
H. pallidus Gebl.	50															
H. laticollis Thoms.	51	●		◐	◐	●	●		◐				◐			
H. strigifrons Thoms.	52	●	◐	●	◐	●	◐				◐		●	◐	●	
H. croaticus Kuw.																
H. dorsalis (Marsh.)																
H. asperatus Rey	53															
H. flavipes F.	54	●	●	●	●	●	●	●	◐	◐	●	●	◐	●	●	●
H. obscurus Muls.	55															
Coelostoma orbiculare (F.)	56	●	◐	◐	◐	●	◐	◐	◐		●	●		◐	◐	
Dactylosternum abdominale (F.)																
Sphaeridium bipustulatum F.	57	●			◐	●	◐	◐	◐							◐
S. lunatum F.	58	●	◐	◐		●								◐	◐	
S. scarabaeoides (L.)	59	●	●	●	◐	●	●	◐	◐		●	●	●	●	◐	◐
Cercyon ustulatus (Preyssl.)	60	●														
C. laminatus Sharp	61															
C. littoralis (Gyll.)	62	●			◐	●	◐	◐	◐		●	●	●		◐	●

	Nn (ø+v)	TR (y+i)	F (v+i)	F (n+ø)	Al	Ab	N	Ka	St	Ta	Sa	Öa	Tb	Sb	Kb	Om	Ok	Ob S	Ob N	Ks	LkW	LkE	Le	Li	Vib	Kr	Lr
31					●	●	●			●															●	●	
32					●	●	●	●	●	●		●	●		●		●		●					●	●	●	
33		◗	◗	●																●	●			●	●	●	●
34						●	●	●	●	●	●	●	●	●	●	●	●	●		?	●				●	●	
35		◖			●	●	●			●																	
36					●			●																	●		
37				●										●			●	●			●	●					●
38					●	●	●	●	●	●	●	●	●	●	●	●	●	●	●						●	●	
39										●																	●
40	◗	●	●	●																			●	●			●
41					●	●	●	●	●	●	●	●	●	●	●	●			●						●		
42																									●		
43					●	●	●			●															●		
44		◗	◗	●			●					●	●		●	●	●			●	●				●	●	●
45																											
46					●																						
47				◗	●	●	●		●	●	●			●		●	●	●			●	●			●	●	●
48																											
49																											
50		◗												●	●		●	●	●		●					●	●
51					●	●	●	●	●	●	●			●	●	●	●			●					●	●	●
52		◗	◗		●	●	●	●	●	●	●	●	●	●	●	●	●				●				●	●	●
53																											
54	◗	●	●	●	●	●	●	●	●	●	●	●	●	●	●	●	●	●	●	●	●				●	●	●
55																											
56					●	●	●	●	●	●	●	●	●	●	●		●		●						●	●	
57					●	●	●	●	●	●	●		●	●											●	●	
58					●	●	●	●	●	●	●	●	●	●	●	●	●	●							●	●	
59		●			●	●	●	●	●	●	●	●	●	●	●	●	●	●							●	●	
60					●	●	●		●	●		●	●												●	●	
61						●	●	●	●	●						●											
62	◗	●		◖	●	●	●	●	●	●															●	●	●

		Germany	G. Britain	SJ	EJ	WJ	NWJ	NEJ	F	LFM	SZ	NWZ	NEZ	B	Sk.	Bl.
Cercyon depressus Stph.	63	●	●	●	●	●		●		●	●	●	●	●	●	●
C. obsoletus (Gyll.)	64	●	●	●	●	●	●		●	●	●	●	●	●	●	
C. impressus (Sturm)	65	●	●	●	●	●	●		●	●	●	●	●	●		●
C. haemorrhoidalis (F.)	66	●	●	●	●	●	●	●	●	●	●	●	●	●	●	●
C. melanocephalus (L.)	67	●	●	●	●	●	●	●	●	●	●	●	●	●	●	●
C. emarginatus Baranowski	68															
C. borealis Baranowski	69															
C. lateralis (Marsh.)	70	●	●	●	●	●	●	●	●	●	●	●	●	●	●	●
C. bifenestratus Küst.	71	●	●	●	●	●	●	●	●	●	●	●	●	●	●	●
C. marinus Thoms.	72	●	●	●	●	●	●	●	●	●	●	●	●	●	●	●
C. unipunctatus (L.)	73	●	●	●	●	●	●	●	●	●	●	●	●	●	●	●
C. quisquilius (L.)	74	●	●	●	●	●		●		●	●	●	●	●	●	●
C. terminatus (Marsh.)	75	●	●	●	●		●	●	●	●		●		●	●	●
C. pygmaeus (Ill.)	76	●	●	●	●	●	●	●	●	●	●	●	●	●	●	●
C. atricapillus (Marsh.)	77	●	●	●	●		●	●	●	●		●		●	●	●
C. granarius Er.	78	●	●		●		●		●	●	●		●			
C. tristis (Ill.)	79	●	●	●	●	●	●	●	●	●	●	●	●	●		
C. convexiusculus Stph.	80	●	●	●	●	●	●	●	●	●	●	●	●	●	●	●
C. sternalis (Sharp)	81	●	●	●	●		●	●	●	●	●	●	●	●	●	●
C. analis (Payk.)	82	●	●	●	●	●	●	●	●	●	●	●	●	●	●	●
Megasternum obscurum (Marsh.)	83	●	●	●	●	●	●	●	●	●	●	●	●	●	●	●
Cryptopleurum subtile Sharp	84	●	●		●	●		●		●	●		●		●	●
C. minutum (F.)	85	●	●	●	●	●	●	●	●	●	●	●	●	●	●	●
C. crenatum (Pz.)	86	●	●	●	●	●	●	●	●	●	●	●	●	●	●	●
Paracymus aeneus (Germ.)	87	●	●	●			●	●	●	●	●		●	●		
Anacaena globulus (Payk.)	88	●	●	●	●	●	●	●	●	●	●	●	●	●	●	●
A. lutescens (Stph.)	89	●	●	●	●	●	●	●	●	●	●	●	●	●	●	●
A. limbata (F.)	90	●	●	●	●	●	●	●	●	●	●	●	●	●	●	●
Hydrobius fuscipes (L.)	91	●	●	●	●	●	●	●	●	●	●	●	●	●	●	●
H. arcticus Kuw.	92															
Limnoxenus niger (Zschach)	93	●	●							●	●					
Laccobius decorus (Gyll.)	94													●		
L. albipes Kuw.		●														
L. minutus (L.)	95	●	●	●	●	●	●	●	●	●	●	●	●	●	●	●
L. cinereus Motsch.		●														
L. biguttatus Gerh.	96	●	●	●	●	●	●	●		●	●	●		●	●	●
L. sinuatus Motsch.	97	●	●		●	●	●	●	●	●	●		●	●	●	●
L. striatulus (F.)	98	●	●	●	●	●	●		●	●	●	●	●	●	●	●

	Hall.	Sm.	Öl.	Gtl.	G. Sand.	Ög.	Vg.	Boh.	Dlsl.	Nrk.	Sdm.	Upl.	Vstm.	Vrm.	Dlr.	Gstr.	Hls.	Med.	Hrj.	Jmt.	Ång.	Vb.	Nb.	Ås. Lpm.	Ly. Lpm.	P. Lpm.	Lu. Lpm.	T. Lpm.
63	●	●	●	●	●		●	●			●	●					●											
64	●	●	●	●		●	●	●			●	●		●	●			●				●	●					
65	●	●	●	●		●	●	●	●	●	●	●	●	●	●	●	●	●	●	●	●	●	●	●	●	●	●	●
66	●	●	●	●	●	●	●	●	●	●	●	●	●	●	●	●	●	●	●	●	●	●		●	●	●	●	●
67	●	●	●	●		●	●	●	●	●	●	●	●	●	●	●	●	●	●	●	●			●	●	●	●	●
68														●	●		●			●	●			●	●		●	
69															●		●			●	●			●	●	●	●	
70	●	●	●	●	●	●	●	●	●		●	●		●	●		●	●		●	●			●	●	●	●	
71		●	●	●	●	●	●	●			●			●			●			●	●			●	●	●	●	
72	●	●	●	●	●	●	●	●	●	●	●	●	●	●	●	●	●	●		●	●			●	●	●	●	
73	●	●	●	●	●	●	●	●	●	●	●	●	●	●	●	●	●	●	●	●	●	●		●	●	●	●	
74	●	●	●	●	●	●	●	●	●	●	●	●	●	●	●	●	●	●		●	●			●	●	●	●	
75	●	●	●	●	●	●	●	●	●	●	●	●	●	●	●	●	●	●		●	●			●	●	●	●	
76	●	●	●	●	●	●	●	●	●	●	●	●	●	●	●	●	●	●		●	●			●	●	●	●	
77	●		●	●			●	●			●			●	●		●	●		●			●					
78																												
79	●	●	●	●	●	●	●	●	●	●	●	●	●	●	●	●	●	●		●	●			●			●	
80	●	●	●	●	●	●	●	●	●	●	●	●	●	●	●	●	●	●		●	●			●			●	
81		●	●	●			●			●			●	●	●													
82	●	●	●	●	●	●	●	●	●	●	●	●	●	●	●	●	●	●	●	●	●	●		●	●	●	●	
83	●	●	●	●	●	●	●	●	●	●	●	●	●	●	●	●	●	●	●	●	●	●		●	●	●	●	
84	●	●		●			●	●				●	●	●	●		●			●	●			●	●	●	●	
85	●	●	●	●		●	●	●			●	●	●	●	●		●	●		●	●			●	●	●	●	
86		●	●	●		●	●	●			●	●	●	●	●		●	●		●	●		●	●	●	●	●	
87																												
88	●	●	●	●	●	●	●	●	●	●	●	●	●	●		●	●	●		●								
89	●	●	●	●	●	●	●	●	●	●	●	●	●	●	●	●	●	●	●	●	●	●	●	●		●		
90																												
91	●	●	●	●	●	●	●	●	●	●	●	●	●	●	●	●	●	●	●	●	●	●	●	●	●	●	●	●
92																							●	●	●	●	●	●
93																												
94		●	●	●			●												●		●	●						
95	●	●	●	●		●	●	●	●	●	●	●	●	●	●	●	●	●	●	●	●	●	●	●	●	●	●	●
96	●		●	●		●	●			●	●	●			●								●					
97			●	●		●	●				?	●	●															
98	●	●		●	●		●	●																				

235

		Ø + AK	HE (s+n)	O (s+n)	B (ø+v)	VE	TE (y+i)	AA (y+i)	VA (y+i)	R (y+i)	HO (y+i)	SF (y+i)	MR (y+i)	ST (y+i)	NT (y+i)	Ns (y+i)
Cercyon depressus Stph.	63															
C. obsoletus (Gyll.)	64															
C. impressus (Sturm)	65															
C. haemorrhoidalis (F.)	66															
C. melanocephalus (L.)	67															
C. emarginatus Baranowski	68															
C. borealis Baranowski	69															
C. lateralis (Marsh.)	70															
C. bifenestratus Küst.	71															
C. marinus Thoms.	72															
C. unipunctatus (L.)	73															
C. quisquilius (L.)	74															
C. terminatus (Marsh.)	75															
C. pygmaeus (Ill.)	76															
C. atricapillus (Marsh.)	77															
C. granarius Er.	78															
C. tristis (Ill.)	79															
C. convexiusculus Stph.	80															
C. sternalis (Sharp)	81															
C. analis (Payk.)	82															
Megasternum obscurum (Marsh.)	83															
Cryptopleurum subtile Sharp	84															
C. minutum (F.)	85															
C. crenatum (Pz.)	86															
Paracymus aeneus (Germ.)	87															
Anacaena globulus (Payk.)	88															
A. lutescens (Stph.)	89															
A. limbata (F.)	90															
Hydrobius fuscipes (L.)	91															
H. arcticus Kuw.	92															
Limnoxenus niger (Zschach)	93															
Laccobius decorus (Gyll.)	94															
L. albipes Kuw.																
L. minutus (L.)	95															
L. cinereus Motsch.																
L. biguttatus Gerh.	96															
L. sinuatus Motsch.	97															
L. striatulus (F.)	98															

					FINLAND																				USSR		
	Nn (ø + v)	TR (y + i)	F (v + i)	F (n + ø)	Al	Ab	N	Ka	St	Ta	Sa	Oa	Tb	Sb	Kb	Om	Ok	Ob S	Ob N	Ks	LkW	LkE	Le	Li	Vib	Kr	Lr
63	◗				●	●	●	●		●															●		
64					●	●	●	●			●														●	●	
65	◗	●	◗		●	●	●	●	●	●	●		●	●	●	●	●	●	●					●	●	●	
66	●	●	◗	◗	●	●	●	●	●	●	●	●	●	●	●	●	●	●	●	●				●	●	●	●
67	●	●	●	●	●	●	●	●	●	●	●	●	●	●	●	●	●	●	●	●	●	●	●	●	●	●	●
68																				●							
69				◗											●					●							
70	◗	◗		◗	●	●	●	●	●	●	●	●	●	●	●	●	●	●	●	●	●				●	●	●
71					●	●	●	●	●	●	●	●	●	●	●			●							●	●	
72		●	◗		●	●	●	●	●	●	●			●		●		●	●						●	●	●
73	◖	●	◗	●	●	●	●	●	●	●	●	●	●	●	●	●	●	●	●	●	●		●	●	●	●	●
74					●	●	●	●	●	●	●	●	●	●	●	●	●	●	●		●				●	●	●
75		◗	●	◖	●	●	●	●	●	●	●		●	●	●	●	●	●			●				●	●	●
76			◗		●	●	●	●	●	●	●	●	●	●	●	●		●	●						●	●	
77		◗			●	●	●	●	●	●	●		●	●	●	●		●							●	●	
78																											
79					●	●	●	●	●	●	●			●		●		●							●	●	●
80	◗			●	●	●	●	●	●	●	●		●	●	●	●	●	●		●					●	●	●
81					●																						
82	◗	●	●	●	●	●	●	●	●	●	●	●	●	●	●	●		●	●	●	●	●		●	●	●	●
83	◗	●	◗	●	●	●	●	●	●	●	●		●	●	●	●			●	●	●				●	●	●
84						●	●			●	●			●						●							
85	●	●	◗	●	●	●	●	●	●	●	●		●	●	●	●	●	●	●	●	●		●	●	●	●	●
86		◗			●	●	●		●	●	●		●	●	●		●	●			●					●	
87																											
88	◗		?	?		●																					
89		◗			●	●	●	●	●	●	●	●	●	●	●	●	●	●	●	●		●			●	●	●
90																											
91	◗	●	◗	●	●	●	●	●	●	●	●	●	●	●	●	●	●	●	●	●	●	●	●	●	●	●	●
92		◗	◗	●															●	●	●		●				●
93																											
94					●	●	●		●			●				●			●								
95		◗			●	●	●	●	●	●	●	●	●	●	●	●	●	●	●	●		●	●	●	●	●	●
96					●	●	●	●		●	●	●		●	●		●								●	●	
97					●		●		?	?																●	
98																										●	

		Germany	G. Britain	SJ	EJ	WJ	NWJ	NEJ	F	LFM	SZ	NWZ	NEZ	B	Sk.	Bl.
Laccobius bipunctatus (F.)	99	●	●	●	●	●	●	●	●	●	●	●	●	●	●	●
L. gracilis Motsch.		●														
Helochares punctatus Sharp	100	●	●	●		●										
H. obscurus (Müll.)	101	●	●	●	●	●		●	●	●	●		●	●		●
Enochrus melanocephalus (Oliv.)	102	●	●	●	●	●	●	●	●	●	●	●	●	●		●
E. ochropterus (Marsh.)	103	●	●	●	●			●	●	●	●		●	●		●
E. fuscipennis (Thoms.)	104	●	●	●	●	●	●	●	●	●	●	●	●		●	●
E. quadripunctatus (Hbst.)	105	●	●	●	●	●			●				●	●	●	●
E. halophilus (Bedel)	106	●	●						●	●	●		●			
E. bicolor (F.)	107	●	●	●	●			●	●	●	●	●	●	●		●
E. testaceus (F.)	108	●	●	●	●	●		●	●	●	●	●	●	●	●	●
E. affinis (Thbg.)	109	●	●	●	●	●	●	●	●	●	●		●	●	●	●
E. isotae Hebauer		●														
E. coarctatus (Gredl.)	110	●	●	●	●	●	●	●	●	●	●		●	●	●	●
Cymbiodyta marginella (F.)	111	●	●	●	●	●	●	●	●	●	●	●	●	●	●	●
Chaetarthria seminulum (Hbst.)	112	●	●	●	●	●	●	●	●	●	●	●	●	●	●	●
Hydrochara caraboides (L.)	113	●	●	●					●	●	●	●	●	●		●
Hydrophilus piceus (L.)	114	●	●	●	●				●	●	●	●	●		●	
H. aterrimus Eschtz.	115	●		●	●				●	●		●	●	●	●	●
Berosus spinosus (Stev.)	116	●	●	●	●	●	●	●	●	●	●		●		●	●
B. signaticollis (Charp.)	117	●	●	●	●			●							●	●
B. luridus (L.)	118	●	●	●	●	●	●	●	●	●	●	●	●	●	●	●

SWEDEN

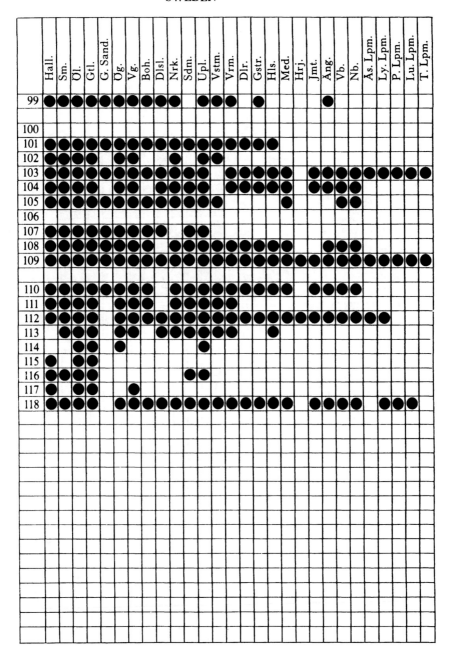

239

Species	No.	Ø+AK	HE (s+n)	O (s+n)	B (ø+v)	VE	TE (y+i)	AA (y+i)	VA (y+i)	R (y+i)	HO (y+i)	SF (y+i)	MR (y+i)	ST (y+i)	NT (y+i)	Ns (y+i)
Laccobius bipunctatus (F.)	99	●		◐	◐	●	●		◐				◑			
L. gracilis Motsch.																
Helochares punctatus Sharp	100															
H. obscurus (Müll.)	101	●	◐		◐	●	◐		◐							
Enochrus melanocephalus (Oliv.)	102	●				●										
E. ochropterus (Marsh.)	103	●	◐	◐		●	◑	◐	◐		●				◑	◑
E. fuscipennis (Thoms.)	104	●	◐	◐	◐	●	●		◐	◐					◑	◑
E. quadripunctatus (Hbst.)	105	◐														
E. halophilus (Bedel)	106															
E. bicolor (F.)	107	●				●		◐	◐	◐	◐		◐			
E. testaceus (F.)	108	●				●										
E. affinis (Thbg.)	109	●	●		◐	●	◐	◐	◐		●		◑	◐	◑	◑
E. isotae Hebauer																
E. coarctatus (Gredl.)	110	●				●		◐								
Cymbiodyta marginella (F.)	111					●				◐						
Chaetarthria seminulum (Hbst.)	112	●	◐	◐	◐	●	●	◐	◐	●	●	◐		●	●	●
Hydrochara caraboides (L.)	113	◐			◐											
Hydrophilus piceus (L.)	114	◑														
H. aterrimus Eschtz.	115															
Berosus spinosus (Stev.)	116	●														
B. signaticollis (Charp.)	117															
B. luridus (L.)	118	●	●		◑	◐	●	●	◐	◐	◐			◐		

Distribution chart

	FINLAND																								USSR		
	Nn (ø+v)	TR (y+i)	F (v+i)	F (n+ø)	Al	Ab	N	Ka	St	Ta	Sa	Öa	Tb	Sb	Kb	Öm	Ök	Öb S	Öb N	Ks	LkW	LkE	Le	Li	Vib	Kr	Lr
99					●	●	●	●	●	●	●		●	●	●										●	●	
100																											
101					●	●	●	●	●	●	●	●	●	●	●	●									●	●	
102					●	●	●	●				●													●	●	
103		◖			●	●	●	●	●	●	●	●	●	●	●	●	●	●	●	●				◖	●	●	●
104					●	●	●	●	●	●	●	●	●	●	●										●	●	●
105						●	●	●	●	●	●	●		●		●									●	●	
106																											
107					●	●	●	●		●	●	●															
108					●	●	●	●	●	●	●	●	●	●	●										●	●	
109		◖			●	●	●	●	●	●	●	●	●	●	●	●	●	●	●	●	●	●			◖	●	●	●
110					●	●	●	●	●	●	●	●			●										●	●	
111					●	●	●	●	●	●	●	●													●	●	
112					●	●	●	●	●	●	●	●	●	●	●		●	●		●					●	●	●
113					●					●															●	●	
114						●				●																	
115						●																			●		
116					●	●	●																				
117																											
118					●	●	●	●	●	●	●	●	●	●	●	●	●			●					●	●	

Literature

Angus, R. B., 1969: Revisional notes on *Helophorus* F. 1.– General introduction and some species resembling *H. minutus* F. – Entomologist's mon. Mag. 105: 1-24.

- 1970a: Revisional notes on *Helophorus* F. 2.– The complex round *H. flavipes* F. – Entomologist's mon. Mag. 106: 129-148.

- 1970b: Revisional notes on *Helophorus* F. 3.– Species resembling *H. strigifrons* Thoms. and some further notes on species resembling *H. minutus* F. – Entomologist's mon. Mag. 106: 238-256.

- 1970c: Revisional studies of East Palearctic and some Nearctic species of *Helophorus* F. (Coleoptera: Hydrophilidae). – Acta zool. hung. 16: 249-290.

- 1970d: A revision of the beetles of the genus *Helophorus* F. (Coleoptera: Hydrophilidae) subgenera *Orphelophorus* d'Orchymont, *Gephelophorus* Sharp and *Meghelophorus* Kuwert. – Acta zool. fenn. 129: 1-62.

- 1971: *Helophorus brevipalpis* Bedel in North America. – Coleopts Bull. 25 (4): 129-130.

- 1973a: Pleistocene *Helophorus* from Borislav and Starunia in the Western Ukraine, with a reinterpretation of M. Lomnicki's species, description of a new Siberian species, and comparison with British Weichselian faunas. – Phil. Trans. R. Soc. B, 265 (869): 299-326.

- 1973b: The habitats, life histories and immature stages of *Helophorus* F. – Trans. R. ent. Soc. Lond. 125: 1-26.

- 1974: Notes of the *Helophorus* species of Fennoscandia and northern Russia. – Notul. ent. 54: 25-32.

- 1977a: A re-evaluation of the taxonomy and distribution of some European species of *Hydrochus* Leach. – Entomologist's mon. Mag. 112: 177-201.

- 1977b: Pale *Helophorus obscurus* Muls. in Worcestershire, with a note on its genetics. – Entomologist's mon. Mag. 112: 201.

- 1982: Separation of two species standing as *Helophorus aquaticus* (L.) (Coleoptera, Hydrophilidae) by banded chromosome analysis. – Syst. Ent. 7: 265-281.

- 1983a: Evolutionary stability since the Pleistocene illustrated by reproductive compatibility between Swedish and Spanish *Helophorus lapponicus* Thomson (Coleoptera, Hydrophilidae). – Biol. J. linn. Soc. 19: 17-25.

- 1983b: Separation of *Helophorus grandis, maritimus* and *occidentalis* sp.n. (Coleoptera, Hydrophilidae) by banded chromosome analysis. – Syst. Ent. 8: 1-13.

Bach, M., 1866: Systematisches Verzeichniss der Käfer Deutschlands, gehörend zur Käferfauna. 64 pp. – Coblenz.

Balfour-Browne, F., 1958: British water beetles. Vol. III. 53 + 210 pp. – London.

Balfour-Browne, J., 1938: A contribution to the study of the Palpicornia. I. – Entomologist's mon. Mag. 74: 102-106.

- 1948: On a new species of *Ochthebius (Asiobates)* of the *bicolon*-group from Finland. – Notul. ent. 28: 95-96.

Bameul, F., 1984: *Enochrus (Methydrus) isotae* Hebauer et espèces voisines dans le sud-ouest de la France (Insecta, Coleoptera, Hydrophilidae). – Bull. Soc. linn. Bordeaux 12: 141-148.

Bangsholt, F., 1981: Femte tillæg til "Fortegnelse over Danmarks biller (Coleoptera)". – Ent. Meddr 48: 49-103.

Baranowski, R., 1976: Några för Sverige nya skalbagger. – Ent. Tidskr. 97: 117-123.

- 1977: Intressanta skalbaggsfynd 2 (Coleoptera). – Ent. Tidskr. 98: 133-140.

- 1978: Intressanta skalbaggsfynd 3 (Coleoptera). – Ent. Tidskr. 99: 53-60.

- 1979: Intressanta skalbaggsfynd 4. – Ent. Tidskr. 100: 71-80.

- 1985: Two new species of *Cercyon* Leach from boreal Sweden (Coleoptera: Hydrophilidae). – Ent. Scand. 15: 341-347.
Bedel, L., 1878: Une espèce d'Hydrophilide encore inédite: *Philydrus halophilus.* – Annls Soc. ent. Fr. (5) 8: 169-170.
- 1881: Faune des Coléoptères du bassin de la Seine (et de ses bassins secondaires) Vol. 1. App. to Annls Soc. ent. Fr. 24 + 360 pp.
Bellstedt, R., 1982: *Hydrochus ignicollis* Motschulsky, 1860 in der DDR (Col., Hydraenidae). – Ent. Nachr. 26: 79-81.
Berge Henegouwen, A. L. van, 1984: Een opmerklijk verschilkenmerk voor de wijfjes van *Hydrochus elongatus* (Schaller) en *H. ignicollis* Motschulsky (Coleoptera: Hydrophilidae) – Ent. Ber., Amst. 44: 33-34.
- 1985: *Enochrus (Methydrus) isotae* Hebauer, nieuw voor Nederland (Coleoptera: Hydrophilidae). – Ent. Ber., Amst. 45: 61-63.
- 1986: Revision of the European species of the genus *Anacaena* Thomson (Coleoptera, Hydrophilidae). – Ent. Scand. 17: 393-407.
Berthold, A. A., 1827: Latreille's natürliche Familien des Thierreiches. (German translation of Latreille: Familles naturelles... 1825). 8 + 602 pp. – Weimar.
Biström, O. & Silfverberg, H., 1983: Additions and corrections to Enumeratio Coleopterorum Fennoscandiae et Daniae. – Notul. ent. 63: 1-9.
Bosc d'Antic, L. A. G., 1791: *Opatrum rufipes* spec. nov. – Bull. Soc. Philomath. Paris 1: 8.
Brullé, A., 1835: Histoire Naturelle des Insectes. Vol. 5, Col. II. 436 pp. – Paris.
Brundin, L., 1934: Die Coleopteren des Torneträskgebietes. 436 pp. – Lund.
Bøving, A. G., & Henriksen, K. L., 1938: The developmental stages of the Danish Hydrophilidae. – Vidensk. Medd. Dansk naturh. Foren. 102: 27-162.
Carr, R., 1984: *Limnebius crinifer* Rey new to Britain, with a revised key to the British *Limnebius* species (Coleoptera: Hydraenidae). – Entomologist's Gaz. 35: 99-102.
Charpentier, T., 1825: Horae Entomologicae, adjectis tabulis novem coloratis. 16 + 255 pp., 7 pl. – Wratislaviae.
Chiesa, A., 1959: Hydrophilidae Europae. 199 pp. – Bologna.
Cuppen, H. P. J. J., 1982: *Hydrochus ignicollis* Motschulsky, nieuw voor Nederland (Coleoptera: Hydraenidae). – Ent. Ber., Amst. 42: 54-55.
- 1983: Een oecologisch onderzoek naar de macrofauna van een temporair kwelmoeras op de Oost-Veluwe. 15 pp. Regionale Milieuraad, Oost-Veluwe.
Cuppen, J., 1981: *Hydraena bohemica* Hrbáček, nieuw voor Belgie en Nederland (Coleoptera, Hydraenidae). – Phegea 9: 61-64.
Cuppen, J. & Cuppen, H. P. J. J., 1982: *Hydraena britteni* Joy new for the Netherlands (Coleoptera: Hydraenidae). – Ent. Ber., Amst. 42: 45-48.
Curtis, J., 1830: British Entomology. Vol. 7. Pl. 290-337. – London.
Derksen, W. & Scheiding, U., 1963-72: Index Litteraturae Entomologicae, Serie 2: Die Welt-Literatur über gesammte Entomologie von 1864 bis 1900. Vol. 1-4. 12 + 697 pp. (vol. 1, 1963); 678 pp. (vol. 2, 1965); 528 pp. (vol. 3, 1968); 482 pp. (vol. 4, 1972). – Berlin.
Duftschmid, C., 1805: Fauna Austriaca... Vol. 1. 311 pp. – Linz u. Leipzig.
Emden, F. I. van, 1956: The *Georyssus* larva – a Hydrophilid. – Proc. R. ent. Soc. Lond. (A) 31: 20-24.
Erichson, W. F., 1837: Die Käfer der Mark Brandenburg 1. 740 pp. – Berlin.
Eriksson, U., 1972: The invertebrate fauna of the Kilpisjärvi area, Finnish Lapland. 11. Haliplidae, Gyrinidae and Hydrophilidae. – Acta Soc. Fauna Flora fenn. 80: 161-164.
Eschscholtz, J. F., 1822: Entomographien. Erste Lieferung. 3 + 128 pp., 2 pl. – Berlin.
Fabricius, J. C., 1775: Systema Entomologicae... 30 + 832 pp. – Flensburgi et Lipsiae.

- 1776: Genera Insectorum... 14 + 310 pp. - Chilonii.
- 1781: Species Insectorum... Vol. 1. 8 + 552 pp. - Hamburgi et Kilii.
- 1787: Mantissa Insectorum... Vol. 1. 20 + 348 pp. - Hafniae.
- 1792: Entomologica Systematica. Vol. 1 (pars 1 & 2). 20 + 330 pp. (pars 1) + 538 pp. (pars 2). - Hafniae.
- 1801: Systema Eleutheratorum... Vol. 1. 24 + 506 pp. - Kiliae.

Fauvel, A., 1887: Rectifications au Catalogus Coleopterorum Europæ et Caucasi. (Suite). - Revue Ent. 6: 75-96.

Forster, J. R., 1771: Novæ species Insectorum. Centuria 1. 8 + 100 pp. - London.

Foster, G. N., 1984: Notes on *Enochrus* subgenus *Methydrus* Rey (Coleoptera: Hydrophilidae), including a species new to Britain. - Entomologist's Gaz. 35: 25-29.

Franck, P. & Sokolowski, K., 1929: Palpicornia und Staphylinoidea des Niederelbe-Gebietes und Schleswig-Holsteins. - Verh. Ver. naturw. Heimatforsch. 21: 47-103.
- 1930: Käfer des Niederelbgebiets und Schleswig-Holsteins IV. Malacodermata, Sternoxia, Fossipedes, Macrodactylia und Brachymera. - Verh. Ver. naturw. Heimatforsch. 22: 79-125.

Freude, H., Harde, K. W. & Lohse, G. A., 1965: Die Käfer Mitteleuropas 1. 214 pp. - Krefeld.

Ganglbauer, L., 1904: Die Käfer von Mitteleuropa. Vol. 4. 1. 286 pp. - Wien.

Gebler, F. A. von, 1830: Bemerkungen über die Insekten Sibiriens, vorzüglich des Altai. - Ledebours Reise T. 2. Pars 2. 228 pp.

Gentili, E., 1974: Descrizione di nuove entità appartmenti al genere *Laccobius* Erichson, 1837 e proposta per un nuovo inquadramento sottogenerico (Coleoptera Palpicornia). - Memorie Mus. civ. Stor. nat. Verona 20: 549-565.

Gentili, E. & Chiesa, A., 1975: Revisione dei *Laccobius* paleartici. - Memorie Soc. ent. ital. 54: 1-188.

Gerhardt, J., 1872: *Hydrobius rottenbergii* n. sp. - Z. Ent. 3: 3-7.
- 1877: Zur Gruppe A der Rottenberg'schen *Laccobius*-Arten. - Z. Ent. 6: 8-27.

Germar, E. F., 1824: Insectorum species novae aut minus cognitae, descriptionibus illustratae. Coleoptera. 24 + 624 pp., 2 pl. - Halae.

Gredler, M. V., 1863: Die Käfer von Tirol. 235 pp. - Bozen.

Gyllenhal, L., 1808: Insecta Suecica. Vol. 1. Pars 1. 572 pp. - Scaris.
- 1827: Insecta Suecica. Vol. 4. 762 pp. - Lipsiae.

Gønget, H., 1961: *Cercyon laminatus* Sharp. (Col., Hydrophilidae). Ny dansk billeart. - Ent. Meddr 31: 74-76.

Hansen, M., 1978: De danske arter af slægten *Hydrochus* Leach, 1817 (Coleoptera, Hydrophilidae) - herunder en ny dansk art. - Ent. Meddr 46: 103-107.
- 1982: Revisional notes on some European *Helochares* Muls. (Coleoptera: Hydrophilidae). - Ent. Scand. 13: 201-211.
- 1983: De danske arter af slægten *Helophorus* Fabricius, 1775 (Coleoptera, Hydrophilidae). - Ent. Meddr 50: 55-76.

Hansen, V., 1931: Biller IX. Vandkærer. - Danmarks Fauna 36: 163 pp. - København.
- 1938: Biller X. Blødvinger, Klannere m.m. - Danmarks Fauna 44: 320 pp. - København.
- 1964: Fortegnelse over Danmarks biller (Coleoptera). - Ent. Meddr 33: 1-507.
- 1970: Tillæg til Fortegnelse over Danmarks biller (Coleoptera). - Ent. Meddr 38: 223-252.

Hanski, I., 1980: The three coexisting species of *Sphaeridium* (Coleoptera, Hydrophilidae). - Annls Ent. Fenn. 46: 39-48.

Hebauer, F., 1981: *Enochrus (Methydrus) isotae* sp. n. - eine neue Hydrophiliden-Art aus Jugoslawien. - Ent. Bl. Biol. Syst. Käfer 77: 137-139.
- 1986: *Ochthebius (Hymenodes) nilssoni* sp.n. from northern Sweden (Coleoptera: Hydraenidae). - Ent. Scand. 17: 359-362.

Heer, O., 1841: Fauna Coleopterorum Helvetica I. 12 + 652 pp. – Turici.

Hellén, W. (Ed.), 1939: Catalogus Coleopterorum Daniae et Fennoscandiae. 7 + 129 pp. – Helsingforsiae.

Herbst, J. F. W., 1793: Natursystem aller bekannten in- und ausländischen Insecten... Vol. 5. 392 pp., 16 pl. – Berlin.

– 1797: Natursystem aller bekannten in- und ausländischen Insecten... Vol. 7. 346 pp., 26 pl. – Berlin.

Hope, F. W., 1838: The Coleopterist's Manual...Vol. 2. Predaceous Land and Water Beetles. 16 + 168 pp., 4 pl. – London.

Horion, A., 1949: Faunistik der Mitteleuropäischen Käfer. II. Palpicornia-Staphylinoidea (ausser Staphylinidae). 388 pp. – Frankfurt a.M.

– 1955: Faunistik der Mitteleuropäischen Käfer. IV. 280 pp., 7 pl. – München.

Horn, W. & Schenkling, S., 1928-29: Index Litteraturae Entomologicae, Serie 1: Die Welt-Literatur über gesamte Entomologie bis inklusive 1863. Vol. 1-4 (vol. 1-3: 1928; vol. 4: 1929). 21 + 1426 pp., 4 pl. – Berlin.

Horn, W. & Kahle, I., 1935-37: Über entomologische Sammlungen, Entomologen & Entomo-Museologi. (Ein Beitrag zur Geschichte der Entomologie). – Repr. from Ent. Beih. Berl.-Dahlem. Vol. 2-4. 6 + 536 pp., 38 pl.

Hrbáček, J., 1951: Revue des espèces du genre *Hydraena* Kug. sur le territoire de la république Tchécoslovaque (Col. Hydroph.). – Čas. čsl. Spol. ent. 48: 201-226.

Huijbregts, J., 1982: De nederlandse soorten van het genus *Cercyon* Leach (Coleoptera: Hydrophilidae). – Zoologische Bijdragen 28: 127-173.

ICZN, 1964: Opinion 710. *Enhydrus* Laporte, 1834 (Insecta, Coleoptera): Validated under the plenary powers. – Bull. zool. Nom. 21: 242-245.

– 1981: Opinion 1178. *Megasternum* Mulsant, 1844 and *Cryptopleurum* Mulsant, 1844 (Insecta, Coleoptera): Type species designated. – Bull. zool. Nom. 38: 114-116.

Ienistea, M.-A., 1968: Die Hydraeniden Rumäniens (Coleoptera, Hydraenidae). – Trav. Mus. Hist. nat. Gr. Antipa: 759-795.

Illiger, J. C. W., 1798: Verzeichniss der Käfer Preussens... 41 + 510 pp. – Halle.

– 1801: Magazin für Insectenkunde. Vol. 1, Heft 1-2. 8 + 260 pp. – Braunschweig.

Janssens, E., 1967: 91. Hydraenidae. Ergebnisse der zoologischen Forschungen von Dr. Z. Kaszab in der Mongolei. (Coleoptera). – Reichenbachia 9 (5): 53-58.

– 1971: A propos d'*Hydraena* s.str. *sternalis* Rey (Col. Hydraenidae). – Bull. Annls Soc. r. ent. Belg. 107: 377-381.

Joy, N. H., 1907: *Hydraena britteni* sp. nov., a new British beetle. – Entomologist's mon. Mag. 43: 79-81.

Kangas, E., 1968a: *Cercyon granarius*, en för Östfennoskandien ny skalbagge. – Notul. ent. 48: 246.

– 1968b: Über *Helophorus walkeri* Sharp und *flavipes* Fabr. (Col., Hydrophilidae) in Finnland. – Annls Ent. Fenn. 34: 38-41.

Kloet, G. S. & Hincks, W. D., 1977: A check list of British Insects. Vol. 11, part 3 (2nd ed.). Hydrophilidae: pp. 12-17. – London.

Knisch, A., 1924. Hydrophilidae. *In* Junk et Schenkling: Coleopterorum Catalogus XIV, pars 79. 306 pp. – Berlin.

Kryger, J. P. & Sønderup, H. P. S., 1940: Biologiske Iagttagelser over 200 Arter af danske Billelarver I. – Ent. Meddr 22: 57-136.

– 1945: Biologiske Iagttagelser over 200 Arter af danske Billelarver II. – Ent. Meddr 24: 175-261.

– 1952: Biologiske Iagttagelser over 200 Arter af danske Billelarver III. – Ent. Meddr 26: 281-349.

Kugelann, J. G., 1794: Verzeichniss der in einigen Gegenden Preussens bis jeztz endeekten Käfer-arten nebst kurzen Nachrichten von denselben (pp. 513-582). *In* Schneider, D. H.: Neuestes Magazin für die Liebhaber der Entomologie. Heft. 5. Pp. 513-640. - Stralsund.

Kuwert, A., 1885: Beiträge zur Kenntniss der Helophoren aus Europa und den angrenzenden Ländern. - Wien. ent. Ztg 4: 229-232, 261-264, 309-312.

- 1886a: Beiträge zur Kenntniss der Helophoren aus Europa und den angrenzenden Ländern.- Wien. ent. Ztg 5: 90-92, 135-139, 169.

- 1886b: General-Uebersicht der Helophorinen Europas und der angrenzenden Gebiete. - Wien. ent. Ztg 5: 221-228, 247-250, 281-285.

- 1887: Uebersicht der europäischen *Ochthebius*-Arten. - Dt. ent. Z. 31: 369-401, pl. 1-4.

- 1888: Generalübersicht der Hydraenen der europäischen Fauna. - Dt. ent. Z. 32: 113-123.

- 1890a: Bestimmungstabellen der europäischen Coleopteren 19. Hydrophilidae. 1. Abteilung: Hydrophilini. 121 pp. - Brünn. Repr. from Verh. naturf. Ver. Brünn 28.

- 1890b: Bestimmungstabellen der europäischen Coleopteren 20. Hydrophilidae. 2. Abteilung: Sphaeridiini und Helophorini. 172 pp. - Brünn. Repr. from Verh. naturf. Ver. Brünn 28.

Laporte, F. L. (Castelnau), 1840: Histoire Naturelle des Animaux Articulés. Vol. 2. Nécrophages - Trimères. 564 pp., pl. 1-38. - Paris.

Latreille, P. A., 1809: Genera Crustaceorum et Insectum secundum ordinem naturalem in. familias disposita, iconibus exemplisque plurimis explicata. Vol. 4. 399 pp. - Parisiis et Argentorat.

- 1810: Considérations générale sur l'ordre natural des animaux composant les classes des Crustacés, des Arachnides et des Insectes. 444 pp. - Paris.

Lawrence, J. F. & Newton, A. F., 1982: Evolution and classification of beetles. - Ann. Rev. Ecol. Syst. 13: 261-290.

Leach, W. E., 1815: Entomology. *In* Brewster, Edinburgh Encyclopaedia. Vol. 9. Pp. 57-172. - Edinburg.

- 1817: The zoological miscellany. Vol. 3. 151 pp., pl. 121-150. - London.

Leiler, T.-E. & Prütz, P., 1977: Nya landskapsfynd av skalbaggar (Coleoptera). - Ent. Tidskr. 98: 95-96.

Lindberg, H., 1943: *Hydrobius*-formernas systematiska ställning. - Notul. ent. 23: 61-62.

Lindroth, C. H., 1935: Die Coleopterenfauna am See Pjeskejaure im Schwedischen Lappland. - Ark. Zool. 28 A (8): 1-60.

- 1957: The principal terms used for male and female genitalia in Coleoptera. - Opusc. ent. 22: 241-256.

- (Ed.), 1960: Catalogus Coleopterorum Fennoscandiae et Daniae. 479 pp. - Lund.

Linnaeus, C., 1758: Systema Naturae... ed. 10. Vol. 1. 2 + 824 pp. - Holmiae.

- 1761: Fauna Suecica. Ed. 2. 578 pp., 2 pl. - Stockholmiae.

- 1775: Dissertatio Entomologica, Bigas Insectorum Sistens... 8 pp., 1 pl. - Upsaliae.

Lohse, G. A., 1971: Palpicornia. *In* Freude, Harde, Lohse: Die Käfer Mitteleuropas III. Pp. 95-156. - Krefeld.

- 1974: Neue und seltene Käfer des Niederelbgebietes und Schleswig-Holsteins. - Bombus 2 (54): 216.

- 1983: Neue und seltene Arten des Niederelbgebietes und Schleswig-Holstein. - Bombus 2 (71): 282-284

Lomnicki, J., 1911: Przeglad wodolubków *(Philydrus)* Polski (Z rysunkiem w tekście) [Synopsis des *Philydrus* de Pologne] - Kosmos, Warsz. 36: 263-273.

Lundberg, S., 1977: Fynd av för Sverige nya skalbaggsarter rapporterade under åren 1974-75 (Coleoptera). - Ent. Tidskr. 98: 7-9.

- 1978: Fynd av för Sverige nya skalbaggsarter rapporterade under åren 1976-77 (Coleoptera). - Ent. Tidskr. 99: 61-63.

- 1980a: Fynd av för Sverige nya skalbaggsarter rapporterade under åren 1978-79. – Ent. Tidskr. 101: 91-93.
- 1980b: För Norrbotten nya skalbaggar under tioårsperioden 1969-78. – Ent. Tidskr. 101: 147-150.
- 1981: Gotska Sandöns skalbaggsfauna – nytillskott och intressanta arter. – Ent. Tidskr. 102: 147-154.
- 1984: Fynd av för Sverige nya skalbaggsarter rapporterade under åren 1982-83. – Ent. Tidskr. 105: 153-154.
Maillard, Y. P., 1968: L'appareil fileur des Coléoptères Hydrophilidae données structurales et fonctionnelles. – Annls Soc. ent. Fr. 4: 503-514.
Marsham, T., 1802: Entomologica Britannica. Vol. 1. Coleoptera. 31 + 548 pp. – London.
Méguignon, A., 1937: Observations sur quelques noms de genre I. *Hydrophilus* ou *Hydrous?* – Bull. Soc. ent. Fr. 42: 53-55.
Meybohm, H ., 1981: Neue und seltene Käfer aus Schleswig-Holstein und dem Niederelbgebiet. – Bombus 2 (68): 269-270.
Motschulsky, V. de, 1849: Coléoptères recus d'un voyage de M. Handschuh dans le midi de l'Espagne énumérés et suivis de notes. – Bull. Soc. Imp. Nat. Moscou 22: 52-163.
- 1853: Hydrocanthares de la Russie,... 15 pp. – Helsingfors.
- 1855: Nouveautés. – Etud. Entom. 4: 77-84.
- 1860: Coléoptères de la Sibérie orientale et en particulier des rives de l'Amour (rapportés par Schrenk, Maack, Ditmar, Voznessenski). – Reisen und Forschungen im Amur-Lande in den Jahren 1854-1856 von Dr. Leopold v. Schrenk 2 (2): 77-257.
Mulsant, E., 1844: Histoire Naturelle des Coléoptères de France. Palpicornia. 7 + 196 pp., 1 pl. (errata et addenda: 197). – Paris.
- 1846: Histoire Naturelle des Coléoptères de France. Sulcicolles et Sécuripalpes (with a suppl. to Palpicornes, Lamellicornes & Longicornes). 280 pp., 1 pl.; suppl.: 26 pp. – Paris.
Müller, O. F., 1764: Fauna Insectorum Fridrichsdalina... 24 + 96 pp. – Hafniae et Lipsiae.
- 1776: Zoologicæ Danicæ Prodromus, seu Animalium, Daniæ et Norvegiæ Indigenarum... 32 + 274 pp. – Havniæ.
Nilsson, A. N., 1982: Aquatic Coleoptera of the northern Swedish Bothnian coast. *In* Müller, K.: Coastal Research in the Gulf of Bothnia. – Monographiae biol. 45: 273-283.
- 1984: The distribution of the aquatic beetle family Hydraenidae (Coleoptera) in Northern Sweden, with an addenda to the Elmidae. – Fauna Norrlandica 4: 1-12.
Nyholm, T., 1952: *Cercyon janssoni* n.sp., eine neue *Cercyon*-Art aus Schweden (Col., Hydrophilidae). – Ent. Tidskr. 73: 207-211.
Olivier, A. G., 1790: Entomologie, ou Histoire Naturelle des Insectes... Vol. 2. 485 pp., 63 pl. – Paris.
- 1792: Encyclopédie méthodique. Dictionnaire des Insectes. Vol. 7. 827 pp. – Paris.
Orchymont, A. d', 1925: Faune des Coléoptères de la Région Lyonnaise. Genre *Helophorus* F. (Hydrophilidae). – Annls Soc. linn. Lyon 72: 111-142.
- 1928: Catalogue of Indian Insects 14 (Palpicornia). 146 pp. – Calcutta.
- 1929a: Notes pour la classification des *Aulacochthebius* (Kuwert) Ganglbauer. – Bull. Annls Soc. ent. Belg. 69: 191-202.
- 1929b: Notes sur quelques *Hydraena* paléarctiques. – Bull. Annls Soc. ent. Belg. 69: 367-386.
- 1932: Zur Kenntnis der Kolbenwasserkäfer (Palpicornia) von Sumatra, Java und Bali. – Arch. Hydrobiol. Supplement Band, IX (Tropische Binnengewasser II): 632-714.
- 1933a: La respiration des Palpicornes aquaticques. – Bull. Annls Soc. ent. Belg. 73: 17-32.
- 1933b: Contribution à l'étude des Palpicornia VIII. – Bull. Annls Soc. ent. Belg. 73: 271-314.
- 1935: Inverta entomologica itineris Hispanici et Maroccani, quod a. 1926 fecerunt Harald et

Håkan Lindberg. XXII. Palpicornia. – Commentat. Biol. 5 (1): 1-22.
- 1936a: Au sujet de la Phylogénie du genre *Hydraena*. – Mém. Mus. r. Hist. nat. Belg. (2) 3: 61-67.
- 1936b: Les *Hydraena* de la Péninsule Ibérique. – Mém. Mus. r. Hist. nat. Belg. (2) 6: 1-48.
- 1936c: Revision des *Coelostoma* (s.str.) non Américains. – Mém. Mus. r. Hist. nat. Belg. (2) 7: 1-38.
- 1937a: Contribution à l'étude des Palpicornia IX. – Bull. Annls Soc. ent. Belg. 77: 213-255.
- 1937b: Changement de noms de genres L' "OPINION" 11. – Bull. Annls Soc. ent. Belg. 77: 423-432.
- 1938a: Contribution à l'étude des Palpicornia XI. – Bull. Annls Soc. ent. Belg. 78: 261-270.
- 1938b: Contribution à l'étude des Palpicornia XII. – Bull. Annls Soc. ent. Belg. 78: 426-438.
- 1939: Contribution à l'étude des Palpicornia XIII. – Bull. Annls Soc. ent. Belg. 79: 357-378.
- 1940: Palpicornia de Chypre. – Mém. Mus. r. Hist. nat. Belg. (2) 19: 1-35.
- 1942a: Palpicornia. Notes diverses et espèces nouvelles III. – Bull. Mus. r. Hist. nat. Belg. 18 (26): 1-20.
- 1942b: Révision du sous-genre *Homalochthebius* Kuwert 1887 du genre *Ochthebius* Leach. – Bull. Mus. r. Hist. nat. Belg. 18 (39): 1-16.
- 1943: Contribution à l'étude du sous-genre *Ochthebius* (s.str.) Kuwert, 1887. – Bull. Mus. r. Hist. nat. Belg. 19 (10): 1-24.
- 1945: Notes nouvelles sur le genre *Limnebius* (Coleoptera Palpicornia Hydraenidae). – Bull. Mus. r. Hist. nat. Belg. 21 (6): 1-24.
Ordish, R. G., 1984: Hydraenidae (Insecta: Coleoptera). – Fauna of New Zealand 6: 56 pp. – Wellington.
Panzer, G. W. F., 1794: Faunae Insectorum Germanicae initia oder Deutschlands Insecten. Hft. 13-24. 16 pp. – Nürnberg.
Paykull, G., 1798: Fauna Suecica, Insecta (Col.) Vol. 1. 8 + 358 pp. – Upsaliae.
Perkins, P. D., 1980: Aquatic beetles of the family Hydraenidae in the Western Hemisphere: Classification, biogeography and inferred phylogeny (Insecta: Coleoptera). – Quaest. Ent. 16: 3-554.
Peyron, E., 1858: Catalogue des Coléoptères des environs de Tarsous (Caramanie) avec la description des espèces nouvelles. – Annls Soc. ent. Fr. (3) 6: 353-434.
Pope, R. D., 1985: *Hydrophilus, Hydrous* and *Hydrochara* (Coleoptera: Hydrophilidae). – Entomologist's mon. Mag. 121: 181-184.
Poppius, B., 1907: Beiträge zur Kentnis der Coleopteren-Fauna des Lena-Thales in Ost-Sibirien. 3. Gyrinidae, Hydrophilidae, Georyssidae, Parnidae, Heteroceridae, Lathridiidae und Scarabaeidae. – Öfvers. finska VetenskSoc. Förh. 49 (1905-06) (2): 1-17.
- 1909: Die Coleopteren-Fauna der Halbinsel Kanin. – Acta Soc. Fauna Flora fenn. 31 (8): 1-58.
Preyssler, J. D. E., 1790: Verzeichniss böhmischer Insekten. Erstes Hundert. 108 pp., 2 pl. – Prag.
Reitter, E., 1885: Neue Coleopteren aus Europa und den angrenzenden Ländern, mit Bemerkungen über bekannte Arten. – Dt. ent. Z. 29: 353-392.
- 1909: Fauna Germanica. Die Käfer des Deutschen Reiches II. 368 pp., 12 pl. – Stuttgart.
Rey, C., 1883: Notices entomologique. – Revue Ent. 2: 84-91.
- 1885a: Descriptions de Coléoptères nouveaux ou peu connus de la tribu des Palpicornes. – Annls Soc. linn. Lyon 31 (1884): 13-32.
- 1885b: Histoire naturelle des Coléoptères de France. (Suite). – Annls Soc. linn. Lyon 31 (1884): 213-396.
- 1886: Histoire naturelle des Coléoptères de France. (Suite). – Annls Soc. linn. Lyon 32 (1885): 1-187, pl. 1-2.
- 1893: Descriptions de deux espèces nouvelles ou peu connues de Coléoptères: *Hydraena* et

Barypithes. – Bull. Soc. ent. Fr. 62: 9-11.
Rossi, P., 1794: Mantissa Insectorum,... Vol. 2. 154 pp. – Pisa.
Rottenberg, A. von, 1874: Revision der Europäischen *Laccobius*-Arten. – Berl. ent. Z. 18: 305-316.
Rydh, I., 1980: Nyfynd af skalbaggar i Blekinge 3. – Ent. Tidskr. 101: 156-157.
Ryker, L. C., 1972: Acoustic behaviour of four sympatric species of water scavenger beetles (Coleoptera, Hydrophilidae, *Tropisternus*). – Occ. Pap. Mus. Zool. Univ. Mich. 666: 1-19.
Sahlberg, J., 1875: Enumeratio Coleopterorum Palpicornium Fenniae... – Notis. Sällsk. Fauna Flora fenn. Förh. 14: 201-227.
Schaller, J. G., 1783: Neue Insecten beschrieben. – Schrift. naturf. Ges. Halle 1: 217-328.
Seidlitz, G. C. M., 1887-1891: Fauna Baltica. Die Käfer (Coleoptera) der deutschen Ostseeprovinzen Russlands. Ed. 2. 10 + LVI + 818 pp. – Königsberg.
Sharp, D., 1869: Description of a new species of *Philhydrus*. – Entomologist's mon. Mag. 5: 240-241.
– 1873: The water beetles of Japan. – Trans. R. ent. Soc. Lond.: 45-67.
– 1884: The water beetles of Japan. – Trans. R. ent. Soc. Lond.: 439-464.
– 1915a: Studies in Helophorini. 1. The genera. – Entomologist's mon. Mag. 51: 2-5.
– 1915b: Studies in Helophorini. 4. – The Empleuri. – Entomologist's mon. Mag. 51: 130-138.
– 1915c: Studies in Helophorini. 6. – *Gephelophorus* and *Meghelophorus*. – Entomologist's mon. Mag. 51: 198-204.
– 1916: Studies in Helophorini. 10. – *Helophorus*. – Entomologist's mon. Mag. 52: 108-112, 125-130, 164-177.
– 1918: On some species hitherto assigned to the genus *Cercyon* (Coleoptera, Hydrophilidae). – Entomologist's mon. Mag. 54: 274-277.
Shatrovskiy, A. G., 1984: A review of Hydrophilids of the genus *Laccobius* Er. (Coleoptera, Hydrophilidae) of the fauna of the USSR (In Russian). – Ent. Obozr. 63 (2): 301-325.
Silfverberg, H. (ed.), 1979: Enumeratio Coleopterorum Fennoscandiae et Daniae. 6 + 79 pp. – Helsingfors.
Smetana, A., 1974: Revision of the genus *Cymbiodyta* Bed. – Mem. ent. Soc. Can. 93: 113 pp.
– 1978: Revision of the subfamily Sphaeridiinae of America north of Mexico. – Mem. ent. Soc. Can. 105: 292 pp.
– 1980: Revision of the genus *Hydrochara* Berth. (Coleoptera: Hydrophilidae). – Mem. ent. Soc. Can. 111: 100 pp.
Solier, A. J. J., 1834: Observations sur la tribu des Hydrophiliens et principalement sur le genre *Hydrophilus* F. – Annls Soc. ent. Fr. 3: 299-318.
Steffan, A. W., 1979: Georissidae. *In* Freude, Harde, Lohse: Die Käfer Mitteleuropas VI. Pp. 294-296. – Krefeld.
Stephens, J. F., 1829: Illustrations of British Entomology... Mandibulata. Vol. 2. 200 pp., pl. 10-15. – London.
– 1832. Illustrations of British Entomology... Mandibulata. Vol. 5. 448 pp., pl. 24-27. – London.
Steven, C. von, 1808: *In* Schoenherr, C. J.: Synonymia Insectorum oder Versuch einer Synonymie aller bisher bekannten Insecten nach Fabricius Syst. Eleutheratorum geordnet,... Vol. 2. (*Spercheus-Cryptocephalus*). 9 + 423 pp, pl. 4. – Stockholm.
Strand, A., 1946: Nord-Norges Coleoptera. – Tromsø Mus. Årsh. 67 (1944), no. 1: 1-629.
– 1965: De nordiske arter av slekten *Helophorus* F. – Norsk. ent. Tidskr. 13: 65-77.
– 1970: Additions and Corrections to the Norwegian Part of Catalogus Coleopterorum Fennoscandiae et Daniae. – Norsk ent. Tidskr. 17: 125-145.
Sturm, J., 1807: Deutschlands Insecten. Vol. 2. 4 + 279 pp., pl. 21-52. – Nürnberg.
– 1836: Deutschlands Insecten. Vol. 10. 108 pp., pl. 216-227. – Nürnberg.

Thomson, C. G., 1853: Oefversigt af de i Sverige funna arter af Familjen Palpicornia. – Öfvers. K. VetenskAkad. Förh. 10: 40-58.
– 1859: Skandinaviens Coleoptera. Vol. 1. 304 pp. – Lund.
– 1867: Skandinaviens Coleoptera. Vol. 9. 407 pp. – Lund.
– 1868: Skandinaviens Coleoptera. Vol. 10. 420 pp. – Lund.
– 1883: (Petites notices entòmologique:) *Hydrobius.* – Annls Soc. ent. Fr. (6) 3: 131.
– 1884: Opuscula Entomologica. Vol. 10. Pp. 937-1040. – Lund.
Thunberg, C. P., 1794: Dissertationes Entomologicae sistens Insecta Suecica. Pars 6. Pp. 73-82. – Upsaliae.
Villa, A., 1833: Coleoptera Europae duplete in Collectione Villa. I. 50 pp. – Milan.
Vogt, H., 1968: *Cercyon*-Studien mit Beschreibung zweier neuer deutscher Arten. – Ent. Bl. Biol. Syst. Käfer 64: 172-191.
West, A., 1940: Fortegnelse over Danmarks biller. – Ent. Meddr 21: 1-664.
Wollaston, T. V., 1854: Insecta Maderensia... 634 pp. – London.
– 1864: Catalogue of the Coleopterous Insects of the Canaries in the collection of the British Museum. 648 pp. – London.
Zaitzev., F. A., 1908: Catalogue des Coléoptères aquatiques des familles Dryopidae, Georyssidae, Cyathoceridae, Heteroceridae et Hydrophilidae. – Trudy russk. ént. Obshch. 38: 323-420.
– 1910: Georyssidae. *In* Junk & Schenkling: Coleopterorum Catalogus XIV, pars 17, pp. 49-52. – Berlin.
– 1938: The palearctic species of the G. *Laccobius* Er. (Coleoptera, Hydrophilidae) [In Russian, with English Summary]. – Trudý zool. Sekt., Baku 2: 109-124.
Ziegler, W., 1983: Neue und seltene Käfer aus Schleswig-Holstein und dem Niederelbgebiet. – Bombus 2 (70): 279-280.
– 1984: Neue und seltene Käfer aus Schleswig-Holstein und dem Niederelbgebiet. – Bombus 2 (72): 287-289.
Zschach, J. J., 1788: Museum N. G. Leskeanum. Pars entomologica ad systema entomologicae Cl. Fabricii ordinata. 136 pp., 3 pl. – Lipsiae.

Index

Synonyms are given in italics. The number in bold refers to the main treatment of the taxon.

252

marinus Thoms. (Cercyon), 140, **151**
maritimus Thoms. (Enochrus), 203
Megahelophorus Kuw., 103
Megasternini, **126**
Megasternum Muls., 126, **158**
Meghelophorus Kuw. (Sharp, emend.), 103
melanocephalus L. (Cercyon), 138, **148**
melanocephalus Oliv. (Enochrus), 196, **198**
Methydrus Rey, **205**
Microlaccobius Gentili, **190**
minimus F. (Ochthebius), 37, **44**
minutum F. (Cryptopleurum), 162, **163**
minutus F. (Helophorus), 99, **111**
minutus L. (Laccobius), 181, **186**
minutus auctt. nec F. (Enochrus), 206
nanulus Rott. (Laccobius), 187
nanus Sturm (Helophorus), 93, **117**
narentinus Reitt. (Ochthebius), 34, **43**
niger Zschach (Limnoxenus), **177**
nigriceps Marsh. (Cercyon), 154
nigrita Germ. (Hydraena), 52, **60**
nilssoni Hebauer (Ochthebius), 37, **45**
nitidus Heer (Anacaena), 171
nitidus Marsh. (Limnebius), 66, **72**
nubilus F. (Helophorus), 90, **100**
obscurum Marsh. (Megasternum), **159**
obscurus Muls. (Helophorus), 96, **124**
obscurus Müll. (Helochares), 193, **194**
obsoletus Gyll. (Cercyon), 139, **146**
ochraceus Steph. (Anacaena), 172
ochropterus Marsh. (Enochrus), 196, **199**
Ochthebius Leach, **33**
Ochthebius Leach, s.str., **46**
orbiculare F. (Coelostoma), **127**
pallidus Gebl. (Helophorus), 93, **119**
palustris Er. (Hydraena), 52, **56**
papposus Muls. (Limnebius), 68, **69**
Paracercyon Seidl., **157**
Paracycreon d'Orch., **143**
Paracymus Thoms., 166, **167**
Paraliocercyon Gglb., 144
Philhydrus auctt., 195, 199
Philydrus Sol., 195, 199
Phothydraena Kuw., **55**
piceus L. (Hydrophilus), **216**
picicrus Thoms. (Hydrobius), 175
picinus Marsh. (Limnebius), 72
picinus auctt. nec Marsh. (Limnebius), 74
porculus Bedel (Helophorus), 90, **100**

Pseudenochrus Łomnicki, 199
pulchella Germ. (Hydraena), 54, **61**
pumilio Er. (Helophorus), 97, **118**
pumilio auctt. nec Er. (Helophorus), 117
punctatus Sharp (Helochares), 192, **193**
pusillus Steph. (Ochthebius), 37, **49**
pygmaea F. (Georissus), 85
pygmaeus Illig. (Cercyon), 138, **154**
quadripunctatus Hbst. (Enochrus), 197, **201**
quadripunctatus auctt. nec Hbst.
    (Enochrus), 200
quisquilius L. (Cercyon), 140, **153**
redtenbacheri Kuw. (Helophorus), 93, **117**
riparia Kugel. (Hydraena), 53, 54, **58**
rottenbergi Gerh. (Hydrobius), 175
rousseti Woll. (Dactylosternum), 128
rufipes Bosc (Helophorus), 90, **99**
sahlbergi Fauv. (Enochrus), 200
scarabaeoides L. (Sphaeridium), 131, **132**
semifulgens Rey (Helophorus), 116
semifulgens auctt. nec Rey (Helophorus), 116
seminulum Hbst. (Chaetarthria), **211**
sibiricus Motsch. (Helophorus), 90, **102**
signaticollis Charp. (Berosus), 220, **222**
sinuatus Motsch. (Laccobius), 182, **188**
Spercheidae, 32, **74**
Spercheus Illig., **75**
Sphaeridiinae, 87, **124**
Sphaeridiini, **125**
Sphaeridium F., 125, **129**
spinosus Stev. (Berosus), 219, **221**
sternalis Rey (Hydraena), 54, **60**
sternalis Sharp (Cercyon), 142, **157**
stockmanni B.-Br. (Ochthebius), 36, **41**
strandi Angus (Helophorus), 91, **106**
striatulus F. (Laccobius), 182, **189**
striatulus auctt. nec F. (Laccobius), 188
strigifrons Thoms. (Helophorus), 97, **120**
subrotundatus auctt. (Hydrobius), 175
subrotundus Steph. (Hydrobius), 175
subsulcatus auctt. nec Rey (Cercyon), 157
subtile Sharp (Cryptopleurum), **162**
terminatus Marsh. (Cercyon), 141, **153**
testacea Curt. (Hydraena), 52, **55**
testaceus F. (Enochrus), 196, **203**
Tricholimnebius Kuw., 68
tristis Illig. (Cercyon), 142, **156**
truncatellus Thunb. (Limnebius), 66, **68**
truncatulus Thoms. (Limnebius), 68, **70**